THE HEALTHY GARDEN

GARDEN

SIMPLE STEPS FOR a GREENER WORLD

Kathleen Norris Brenzel and Mary-Kate Mackey

ABRAMS, NEW YORK

Contents

8 **Introduction**

HEALTHY GARDEN

15 **What Makes a Garden Healthy?**

18 **Design a Healthy Garden**
19 *Gathering of Gardeners:* **Juliet Sargeant**
28 *Gathering of Gardeners:* **Darcy Daniels**
40 *Gathering of Gardeners:* **Debra Lee Baldwin**

43 **Choosing Plants for Healthy Gardens**
52 *Gathering of Gardeners:* **Nancy Buley**
59 *Gathering of Gardeners:* **Marietta and Ernie O'Byrne**
66 *Gathering of Gardeners:* **Janet Sluis**
74 *Gathering of Gardeners:* **Melinda Myers**
84 *Gathering of Gardeners:* **Joe Lamp'l**
90 *Gathering of Gardeners:* **Bob Lilly**

HEALTHY YOU

102 The Health Benefits of Gardens

104 Why Grow Your Own Food?
112 *Gathering of Gardeners:* **Brie Arthur**
122 *Gathering of Gardeners:* **Pat Munts**
133 *Gathering of Gardeners:* **Valerie Rice**

144 Exercise Benefits of Gardens
148 *Gathering of Gardeners:* **Toni Gattone**

151 Gardening for Tranquility
160 *Gathering of Gardeners:* **Perla Sofía Curbelo-Santiago**
170 *Gathering of Gardeners:* **Leslie Bennett**

HEALTHY PLANET

175 Make a Difference Through Gardening

178 *Gathering of Gardeners:* **Teresa Speight**
184 *Gathering of Gardeners:* **Abra Lee**
188 *Gathering of Gardeners:* **Brent Green**
192 *Gathering of Gardeners:* **Ron Vanderhoff**

198 **Resources**
202 **Acknowledgments**
203 **Index**
207 **Photograph Credits**

Introduction

Today, more than ever, we are understanding the many ways in which we humans are connected across countries and continents. And we're noticing how we share this blue planet—its air, water, and land, as well as its complex ecosystems and resources—with all other species. As John Muir observed, "When we try to pick out anything by itself, we find it hitched to everything else in the Universe."

Health is about balance. It occurs when all parts of anything, from cellular structures to stars, live together in harmony and ease. All living things seek health. The word comes from the Middle English for *whole*.

Gardeners today are in a good position to make a difference in the world. You can choose plants that don't gulp water and other precious resources. Grow your own food. You can compost debris, instead of taking it to already-crowded landfills. Use salvaged materials, such as broken concrete, in beautiful new ways. Plant trees, flowers, and berried shrubs to help feed and shelter wildlife—whose natural habitats are fast disappearing as human populations continue to spread into previously pristine lands. Healthy gardens can contribute to the health of the natural world—its streams, bays, beaches, and other watersheds, as well as its marshes, meadows, grasslands, deserts, and forests. Ultimately, the greater the number of healthy backyard ecosystems out there, the more we humans can contribute to the health of the planet we call home. *The Healthy Garden* shows you how.

This book is divided into three sections. In Part One, "Healthy Garden," you will learn how to design a sustainable garden—whether you're a seasoned gardener, a beginning gardener, or a would-be gardener. And how to make the right choices, so that every organism in your garden will work together—soil, plants, location. The bonus? Your garden life will become easier and more fun. In Part Two, "Healthy You," you will discover how your garden supports your own well-being, by providing healthy crops for your table, encouraging you to exercise, or calming your mind. In Part Three, "Healthy Planet," you'll see how people like you have joined with others to put their garden knowledge to work for the health of their communities and the growing world beyond.

Plants invite wildlife to your garden. A mockingbird finds lunch on a shrubby cotoneaster laden with berries.

This book includes the ideas and opinions of more than sixty horticultural experts, whose observations are sprinkled throughout its pages. You'll also find special features entitled "The Gathering of Gardeners." These are nineteen short interviews with professionals and passionate plant people, who offer their lively insights and practical advice on a variety of subjects.

All these experts will help you make your own choices for earth-friendly gardening. Some folks you may know; others you'll meet for the first time. Think of them as your fellow gardeners, the ones you might ask to sit down with you on your balcony or in your backyard, to talk about your plants, your dreams and rewards, over glasses of lemonade.

The Healthy Garden invites you to discover why the humble act of gardening may the most important endeavor you can take on, not only in your own backyard for your own well-being, but for the health of the world beyond your fences, now and in the future.

HEALTHY GARDEN

"We must come to know our essential connection to the wilder earth, because it is here, in the activity of our daily lives, that we most surely affect this earth, for good or ill."

—Lyanda Lynn Haupt, from *Crow Planet: Essential Wisdom from the Urban Wilderness*

Previous page: **Borrow from the surrounding landscape.** If you're new to an area and about to install a garden, first check out your region's most iconic plants to help inform your choices. For this garden in Palm Springs, California, the designer selected the region's two iconic palms: stout-trunked Mexican fan palm (*Washingtonia robusta*), and tall, slender California fan palms (*Washingtonia filifera*), which grow naturally near springs and moist spots in the area. Companion plants include a scattering of shapely golden barrel cacti (*Echinocactus grusonii*) and blue agaves. The smooth white boulders among them are properly nestled into the soil, not positioned on top of the ground, with stone mulch around them to complete the natural look.

Right: **Mimic the natural surroundings.** On the edge of wild land, take your planting cues from native vegetation nearby. That way, you'll have a better idea of what plants will thrive without fuss in your garden. The owners of this home, near California's central coast, chose to cover their nearly bare hillside property with plants that are native to this region. These include live oak trees (*Quercus agrifolia*) for shade, an understory planting of the native shrubs manzanita (*Arctostaphylos*) and coffeeberry (*Frangula californica*), and a sedge meadow (*Carex pansa*).

What Makes a Garden Healthy?

A lovingly tended garden can be a source of joy at any size, whether it's a few pots on a porch, a suburban lot bordered by trees and shrubs, or many sprawling acres. It connects you and your family and friends to the natural world outside your door. And, like wild ecosystems everywhere, a garden reaches its fullest potential when it is healthy. So what *is* a healthy garden? How do you make a healthy garden? And why do healthy gardens matter now more than ever?

Healthy gardens touch the land lightly. They work *with* nature, not against her. They are designed to replicate or echo native plant communities. And, as in the natural places, a healthy garden includes a diverse collection of plants—trees, shrubs, vines, grasses, perennials, and annuals—native or non-native, or a mix of both. All are well-suited to the weather, soils, and exposure in your area.

Such diverse plantings also attract wildlife. Bees buzz about transferring pollen from flower to flower. Butterflies flit among the blooms, sipping nectar. Birds chatter in the trees, and beneficial insects chow down on aphids or whiteflies. Underfoot, earthworms wriggle their way through your soil. All these creatures are great allies for you; they can help keep your garden in balance. A vital garden has little need of petrochemicals or excessive water.

Healthy gardens come in all shapes and sizes, with different levels of care. They can be wild, with native plants allowed to grow untended except for pathways winding through them, or meticulously cared for, with each plant chosen for its design relationship to the others. They can be a collection of containers or a series of raised beds for growing food. But when it comes to care, they all have a few things in common.

Healthy gardens rarely, if ever, need chemicals, including insecticides. Instead, these gardens rely on natural methods to control any insect pests or plant diseases that may appear. Even sprays labeled organic are used with caution, because they are not pest-specific. A spray you choose to kill aphids will kill beneficial insects as well. Better to hose or wipe the aphids off and concentrate on improving your soil with compost and mulch, because in most cases, pests go for the least healthy plants first. Plants in healthy gardens are fed as needed, using organic fertilizers. Trees and shrubs are pruned with clean, sharp, disease-free clippers and pruners.

Design a Healthy Garden

You don't need to tear out everything you have and start over in order to move toward a healthier garden. You can build sustainability into your landscape gradually, in small steps. In fact, the best start you can make is to observe what's there before picking up a shovel. If you have moved to a garden space that's unfamiliar, get to know your land. Observe it to find out how existing plants might reappear or perform through the seasons. Who knows? Hidden treasures may be planted all around you. From that patch of bare dirt, for example, tiny snowdrop bulbs may push up in spring. Or that tree you thought was so boring and green might turn glorious shades of red and orange in the fall.

It's also a healthy practice to recycle or upcycle any plants or other gardening equipment you no longer want. Perhaps a ratty collection of abandoned plastic pots would brighten your balcony after a coat of colorful paint. Or you might want to recycle an ugly concrete pathway by breaking it up and building a low retaining wall with the pieces.

You could start a compost pile to improve your soil and then gradually dig it in for new plantings or use it as mulch around plants. Replace an unused lawn with a meadow of mixed grasses and perhaps a scattering of wildflowers. Or tuck in a few tough evergreen shrubs that produce fruits or berries to attract wildlife. Whether you're making changes slowly or considering a complete garden overhaul, plan ahead. It's a less expensive alternative—and more fun—than impulse buying.

Juliet Sargeant

GARDEN HEALTH—IT'S A BALANCING ACT

"Achieving a balance between people and their landscape." That's how award-winning landscape designer Juliet Sargeant describes a healthy garden. "People feel comfortable in their space—engaged, enthralled," she says. "The garden is a positive place to be."

When that experience is not positive, a designer like Juliet can step in to help. "A lot of people are scared, afraid they'll kill a plant," she says. Or they worry they won't be able to maintain a whole garden. On the other hand, Juliet observes, "There are others who want complete control, as if the outdoor area should be as tidy as the kitchen."

Juliet's goal is to engage people with the natural world in a way that brings satisfying results for both the gardener and the garden. She says, "The garden is the bridge between the client and nature."

Juliet knows about health of all kinds, having shifted her profession from medicine into horticulture more than twenty-five years ago. Surprisingly, the two disparate fields have a lot in common. "Similar skills are required," she says. "Problem solving, bringing the patient on board, and finding a solution together."

So, how can you find a balance with your garden? Here's Juliet's advice:

Identify your garden's problem areas. Juliet tries to understand what her clients dislike about their garden, and then she focuses on solutions. You can do the same. Perhaps you're overwhelmed by weeds, or feeling guilty because those container plants keep dying. If you identify what's bothering you, you can figure out what to do. Often, plant choices solve the problem. If Juliet discovers her client prefers a tidy controlled garden, she takes self-seeding plants off the table. "There are plenty of plants that behave themselves," she notes.

A garden welcomes the family. The Ritchie family enjoys exploring their garden designed by Juliet Sargeant in Sussex, England. "I take a people-centered approach to problem solving, so they can better engage, get to see more of nature," Juliet says. "I try to find out where they are and where they want to be."

Gather garden ideas from multiple sources.
Clipping pages from garden magazines has been a time-honored method, and now Pinterest boards are popular. "That's great," Juliet says, "but a little restricting." She suggests that you go beyond inspirational images. "Let your imagination fly." Juliet suggests looking at your preferences in art or music. Perhaps a lover of classical music would like more formal lines in a garden. Someone whose taste in music runs to the rowdy might connect with an exuberant garden filled with loosely flowing grasses and spikey upright accents.

Make it yours. As Juliet reminds us, "It's *your* personality that can come through in your garden, not the designer's." Creating a design that suits you has powerful benefits. "When you're committed to making your own space a reflection of your own interests, desires, and style," Juliet says, "you'll start to see the natural world beyond your garden in a different way."

Revisit your choices. With her clients' gardens, Juliet says, "I can come back and see how the design is working. Checking back is really important. What's working? What isn't? Be willing to keep adjusting." You can check in with your garden every few seasons to assess your decisions. Making changes doesn't mean you've failed. It's a critical part of learning how to be in a healthy relationship with your garden.

Let your garden give back to you.
Strengthening your ties to the world of plants helps you achieve balance—and health. One of Juliet's first jobs as a designer was a tiny back garden she created for a psychiatrist. The doctor later told Juliet that when she'd had a horrible day at work, she'd come into her house, drop her bag, and go straight out the back door into the garden. "It made me think—here's what I can do," Juliet says. "Garden design is something that enhances people's lives."

Juliet Sargeant is the winner of numerous garden design awards, including a gold medal and a People's Choice Award for the Modern Slavery Garden, at the 2016 RHS Chelsea Flower Show. She is the founder of the Sussex Garden School. Find her work at her website, www.julietsargeant.com.

Page 16: **Embrace existing natives.** Before you re-landscape an existing garden, take stock of any trees or other mature plants that should be saved. This giant saguaro (*Carnegiea gigantea*) is an example. Saguaros are native to Arizona's Sonoran Desert, and they grow only about a foot taller in ten years. They are also protected by law (buy only nursery-grown plants). The owners of this Tucson property left the mature saguaros in place as focal points, then set smaller golden barrel cacti and agaves around them to anchor the composition.

Opposite: **Set off a star.** When a plant looks good all year, it's worth showing off. In this south Florida garden designed by landscape architect Raymond Jungles, that accent plant is a red-leafed bromeliad (*Aechmea blanchitiana*), shown here gracing a log. Its central location adds drama and a touch of heat to an otherwise cool green palette. A date palm fans out its fronds in foreground; the tall, stout-trunked trees in back are baobab trees (*Adansonia digitata*), native to Australia, Madagascar, and parts of Africa.

Embrace Your Place

Ideally, your garden should suit where you live, whether your property sits on the edge of wild land, a well-defined lot in the suburbs, or in the city. If possible, look to natural plant communities nearby and learn from them. Every habitat, whether beach, mountain, plains, desert, or jungle, suggests design lessons. Check out the plants that populate the parks and wild lands near where you live. This research can be fun—especially if it happens during a hike. If the landscape is unfamiliar, download a plant-identifier phone app or carry a field guide. Note which plants grow and thrive there naturally, with no help from a gardener. When the time comes to make your own garden, allow what you've learned to guide your plant choices. Look for those signature plants that are easy to grow where you live.

For example, a garden in the Pacific Northwest might have a woodland feel, much like its surroundings, with conifers, vine maples, and signature rhododendrons rising above a carpet of low ferns. In a high mountain garden in Colorado, drifts of blue columbines could fringe a grassy meadow, as they would in the wild. In California's chaparral country, beautiful natives have evolved over millions of years to withstand droughts and to set new seed following wild-fires. Shrubs such as blue-flowered ceanothus, golden mimulus, and toyon make beautiful and unthirsty garden plants.

Author, speaker, and self-described plant geek C. L. Fornari lives on Cape Cod. "Our signature plant here is the blue mophead hydrangea," she says. In fact, gardeners on the Cape are so proud of what they grow, they celebrate with an annual Hydrangea Festival.

In southern gardens, the native flame azalea and many forms of magnolias are stalwarts. For privacy screening in Florida, homeowners rely on the ubiquitous yew pine (*Podocarpus macrophyllus*) and the areca palm (*Dypsis lutescens*). Sylvia Gordon, owner of the Florida wholesale nursery Landscape by Sylvia Gordon, says that *Clusia rosea*, a plant used in tall hedges all over Miami, is sometimes called the signature or autograph tree. "Carve your name in the leaf, and the carving will turn to golden brown and remain visible as long as the leaf is alive," Sylvia says.

Even if you wish to create a highly stylized garden, such as an English cottage display or a Japanese garden, you can still achieve a robust and sustainable plant community if you choose plants that are well-suited to your location and soil. That may mean making substitutions, such as finding a tougher, hardier boxwood to grow in your colder climate than the ones commonly found in milder English gardens.

Celebrate color. Add a bright burst of color among otherwise quiet, dry-climate plants. In this garden, orange cannas, yellow-flowered gloriosa daisies (*Rudbeckia*), and red and yellow kangaroo paws (*Anigozanthos*) add the vivid hues. The planting is organized around a cooling fountain, made from an urn, with a purple coneflower (*Echinacea purpurea*) brushing its side. This bed attracts birds, bees, and butterflies. Design by Sherry Merciari, Merciari Designs.

Frame a view. If the view from your backyard is as breathtaking as this one in Connecticut, keep your plants low so you can use and enjoy that slope. In this garden, designed by Larry Weaner Associates, red-flowered monardas, purple coneflowers, and assorted wildflowers grow in the foreground, while a meadow carpets the gentle slope leading to the lake.

Plant a transition zone. Patio trees and a small backyard lawn give way to a prairie garden with fall-flowering perennials and mounding grasses, which then transitions to distant views of Colorado hills. Design by Lauren Springer Ogden.

Gathering of Gardeners

Darcy Daniels

PLANTING FOR NOW AND THE FUTURE

Garden health depends on all plants having the room they'll need to grow to their fullest potential without crowding. But how do you set plants far enough apart to allow for future growth, yet still be able to enjoy a lovely, lush show now?

Portland, Oregon, garden designer Darcy Daniels has the answer. When faced with this tricky problem in her clients' gardens and in her own, Darcy divides plants into two categories—*Forever* and *For Now*. Follow her advice, and you'll always give plants their future growing space.

Use Forever plants to provide structure. These are your garden's most important trees and shrubs with specific jobs to do, such as providing shade over a patio, or blocking the view of a neighboring house. Darcy suggests choosing plants that will also add seasonal interest. She's fond of trees

Side yard strategy. Darcy's Forever and For Now plantings blend seamlessly in this narrow Portland, Oregon, side yard. The structural Japanese maple (*Acer palmatum* 'Sango Kaku') at the gate, the twisting upward arms of an evergreen conifer (*Chamaecyparis lawsoniana* 'Wissel's Saguaro'), and the climbing hydrangea (*Hydrangea anomala subs. petiolaris*) on the opposite wall will remain as Forever plants. For Now fillers flank the flagstone pathway, including a mix of golden *Hakonechloa macra* 'Aureola', red autumn fern (*Dryopteris* 'Brilliance'), and red *Heucherella* 'Sweet Tea'. These can be replaced as the trees and vine grow.

like the orangebark stewartia (*Stewartia monadelpha*), which shows off its peeling reddish-brown bark when the leaves are gone in winter.

Research the eventual size. Before buying a long-lived (and expensive) tree or shrub, consult a reliable website or plant encyclopedia to confirm how tall and wide it is likely to grow. For trees, most nursery plant tags will only indicate the size after ten years of growth. That could be reported as fifteen to twenty feet, whereas the mature height might be more like seventy-five feet. Always check.

Measure the space your plants will be filling. Keep in mind that the cute little tree you've just bought in a seven-gallon can could eventually develop a canopy with a fifteen-foot spread. To visualize the eventual size, it's sometimes helpful to put up tall poles, or a wide umbrella. Use a tape measure to make sure your tree will fit. Forever plants are often slow growing, so patience, Darcy counsels, is required. But you don't have to live for years with those bare spaces in between.

Enjoy For Now plants as fillers. "These are the ones I have fun with," Darcy says. With For Now plants, you could choose those that need full sun, until the burgeoning shade of that Forever tree or sprawling shrub takes over. Then, new shade plants can take their place. In sunny spots, try short-lived perennials, with two-to-five-year life spans. For example, Darcy combines blanket flower (*Gaillardia grandiflora* 'Arizona Red Shades') and orange yarrow (*Achillea millefolium*

'Terracotta'). Or fill the open space with biennials, which put out foliage the first year, flowering stalks the next, and then disappear after blooming; silver leaf mullein (*Verbascum bombyciferum* 'Arctic Summer') is one pretty example. Or go for your favorite long-blooming annuals to cover the ground and bring you masses of flowers and a design repetition throughout your garden.

Prepare for change. "Be willing to remove For Now plants once they're no longer useful, or have outgrown their space," says Darcy. "Give yourself permission to make a mess." And don't feel guilty about garden editing. "I don't mind tearing plants out, even the big ones," she says, "because that was part of the original plan."

Darcy Daniels is the owner of Bloomtown Gardens and creator of the garden teaching website eGardenGo.com.

Dream Big—Then Plan

While still in the planning stages, explore what you want your garden to do for you, and what *you* want to do in your garden. Even if at this point you're only caring for three houseplants on a sunny windowsill, go ahead and imagine what your future garden might be. That way, when the chance comes for a bigger bit of earth, you'll already know what kind of garden you want.

Naturally, you'll discard many ideas along the way, but why not start with the garden of your dreams? Brainstorm; talk with others who live with you. Search online for inspiration and start a wish list. Notice which designs and features attract you. Flowers that create bright contrasts of color and form? Elegant arrangements of shrubs, paired to play up their shapes and foliage colors? All flowers? No flowers? That Instagram picture of people sitting around a dining table under strings of vintage light bulbs? Ask yourself which features appeal most. Together, these inspirations might make up your future garden's themes. This exploration of garden ideas is the time for playfulness as much as practicalities.

Start small. Perhaps you simply need a container or two that will flank your entryway and welcome guests. Which way does your doorway face? That will affect your plant choices. Consider maintenance. Would you like to replant those containers seasonally? Or would you want a collection of plants that provides interest year-round? These are questions you can also ask yourself when thinking about beds and borders, or a whole garden makeover. Even if you decide to work with a professional designer or contractor, you'll need to have clear ideas about what you want from your garden.

On paper, dream big, no matter how tiny your space. Even on a postage-stamp plot, you can achieve your goals and still have room for a patio, fire pit, play area, or other amenities for outdoor living. Figure out how much space the outdoor furnishings you're considering will take up. As you'll find out, with any new garden project, a tape measure can be your new best friend.

Vanessa Nagel Gardner, an author and award-winning landscape designer, says, "I've had clients who would present me with a good list of needs for a space—perhaps they wanted a little chair and table for a garden retreat area. But they would always forget about circulation—how am I going to move around in this space? The area to maneuver can take up anywhere from 20 to 60 percent of the actual space." For instance, that small chair may only be two feet by two feet, but when someone sits in it, the space the chair occupies has now doubled. "You've

Opposite: **Invite butterflies.** Choose the right plants, and butterflies will find your garden. Here, pink-flowered *Echinacea* 'Dixie Blaze' mingles with *Agastache* 'Blue Boa' in Terra Nova Nursery's test garden in Canby, Oregon. The yellow and red blooms in front are *Coreopsis verticillata* 'Route 66'. Other good choices for attracting butterflies include bleeding heart (*Dicentra*); bishop's lace (*Ammi majus*); butterfly weed (*Asclepias tuberosa*); common yarrow (*Achillea millefolium*); and hollyhock (*Alcea rosea*).

Overleaf: **Grow what you love.** In winter, this garden, designed by Richard Hartlage/Land Morphology, looks quiet and frosty. But in summer, golden blooms of *Rudbeckia fulgida* 'Deamii' carpet the landscape. All species of these vibrant daisy-flowered perennials attract beneficial insects, birds, and butterflies.

got to be able to get around, and other people will need to get around you while you're there," Vanessa says. "Go back and revisit the function to make sure you've allowed enough room."

When you're designating how different areas of your garden will work, don't forget to leave some small space for wildlife. Along with the plants that will provide shelter and food, you can create a low brush pile in a corner or put in some billowy shrubs that can offer shelter and nesting spots for a wide array of birds and insects. Plan to include water as well, and your garden can support far more life than you know.

The Great Lawn Debate

Much has been written about the thirsty, overfed, over-clipped monoculture that makes up lawns. And yet, turf grass is still the only practical and natural outdoor play surface. It also gives the garden an eye-resting patch of green amid a riot of plants.

So, should you have a lawn? It depends on many factors. First, it depends on what part of the country you live in. In areas where year-round rainfall is the norm, lawns make more sense, and are often more expansive than in other parts of the country with scant rainfall. It's a relatively easy garden chore, though mowing them does take gas or electricity. But all lawns, whether big or small, take a minimum of an inch of water a week to keep grass green. The larger your lawn, the more water it needs. In the arid West, where water is such a precious commodity, consider losing the lawn altogether or substituting a swath of less-thirsty groundcover to provide the calming green you need in your garden.

Second, it depends how you want to use your lawn. For backyard games? If so, what's the smallest it could be? Do you see it as the place for a dining table? A small stone patio would serve your purposes better, keeping chair legs from sinking into the soil. If you envision your lawn as a pathway to get from one area of your garden to another, choose other materials, like bark or gravel. And don't bother with lawns on hills. The hassle of mowing a slope makes turf a lousy candidate for covering them. Low-growing easy-care shrubs would serve you better on hillsides.

Third, it depends on how much time you want to spend. Any lawn upkeep takes time. The more lawn you have, the more time you'll spend. In Salt Lake City, Utah, the folks who run the Jordan Valley Water Conservancy District discovered they could encourage people to shrink their lawns by focusing on time-saving design—a series of steps that resulted in

Create the illusion of space. How can you make a small suburban lot look and feel larger? By lifting plants to add height, and then arranging pathways diagonally. In this California garden, paths covered with crushed rock connect raised beds edged with stone. Drought tolerant plants fill the beds; they include purple flowered lantana, showy milkweed (*Asclepias speciosa*) and strawberry tree (*Arbutus unedo*).

Grow a meadow. If you want to remove your lawn and do away with regular mowing, consider planting a meadow, as the owners of this green expanse did in their backyard in Beaverton, Oregon. According to the owners, after letting their lawn "go feral," they contacted landscape designer Lori Scott and explained their vision for a meadow. It was important to them that the land under their stewardship "be clean and sustainable." They got what they wanted. Now, their tall grasses shimmer in sunlight and dance in the breezes; birds fly about, "tweeting like mad." A pair of metal birds, designed by artist Tim Foertsch, twirl in the breeze above the grasses. The only mowing that needs to be done now is the path through the meadow.

less garden worktime. And having less lawn, says Cynthia Bee, their outreach coordinator, "improves water efficiency without destroying the visual character of the neighborhoods."

Fourth, it depends on how you feel about a perfectly clipped swath of a single plant of upright green blades. That's a monoculture, and it's never found in nature. Monocultures are supported by fertilizers, and possibly weed-killing chemicals, both of which can harm the beneficial critters that live in the soil on your property. Rain or sprinklers can wash those same chemical poisons into the ground beyond your yard. That runoff then leaches into streams, rivers, and oceans. Can you change your idea of what's perfect? Could you lower your expectations to encourage a healthier lawn that is a mix of grasses and ground-hugging plants, like daisies or clover? A mixed lawn will stay colorful longer and make less work for you. Perhaps you can lobby to persuade your homeowner association to lower their expectations too. There's nothing pretty in the competition to have the greenest lawn. Do less and nature will do more for you.

A Word About Water

If you live in the driest parts of western America, such as California, Colorado, Arizona, or New Mexico, you most likely have experienced winters with little or no rainfall, followed by periods of prolonged summer drought. Such conditions call for thoughtfully designed gardens that focus on water conservation, with unthirsty shrubs and trees, small lawns or permeable paving used as lawn replacements, and dry streambeds or sumps for channeling and storing rainwater to irrigate plants.

Taking such steps can really make a difference in your water use. That's what Southern California landscape designer Susanne Jett discovered after conducting a study of two small front yards, each the same size, but with one planted with native plants and the other with lawn. Over the study's nine-year period, Jett noted that the lawn-and-sprinklers garden used 703,013 gallons of water, while the native-and-drip garden used considerably less—130,438 gallons—saving more than half a million gallons of water. Another plus: Birds, bees, and butterflies preferred the native garden's habitat over the expanse of lawn.

Debra Lee Baldwin

GARDEN WHERE YOU'RE PLANTED
"I've always loved plants," says author, artist, and photojournalist Debra Lee Baldwin, who grew up on an avocado ranch in San Diego County, California. Still, she admits it took time and lots of experimenting for her to discover, understand, and celebrate the plants that would eventually inspire her life's work: succulents. Today, her continuing fascination with these plants has resulted in hundreds of watercolor paintings, three books on the subject, and social media platforms with a worldwide following.

Why succulents? That's what would grow on the land surrounding her home. Located in the rolling brown hills of eastern San Diego County, her rocky half-acre lot slopes steeply in places. It's studded with mature native oak trees, so it's shady in patches, but hot and sunny in other places. The soil is heavy clay.

So Debra began experimenting with plants, including perennials, jade plants, and roses. Then one winter when her roses were looking especially tired, she visited a succulent garden farther up the California coast. There she discovered a riot of color, mostly from succulents, with some sending up tall

Right plants flourish in the right place. In the sunny front yard, small succulents roam around larger ones against a backdrop of a golden euonymus shrub. Large succulents include (left to right): a pointy-tipped *Agave* 'Blue Glow', striped *Agave Americana* 'Mediopicta Alba', and orange-tinged *Aloe thraskii*. Rambling in front are (left to right): green aeoniums, a golden barrel cactus, two blue chalk sticks (taller *Senecio mandraliscae* and shorter *Senecio serpens*) and compact *Agave* 'Cream Spike'.

spires of yellow and orange blooms, vivid against the blue sky. It was an aha moment. "I redesigned my garden to accommodate them," she says. "Now, I'm growing hundreds of varieties. It's a collector's garden, not designed." And that suits her just fine.

What does she find so striking, so enticing about these plants? "I appreciate how light hits the leaves, the plants' strong geometry, the patterns they produce, of light and shade," Debra says. "I can't wait to capture all that on film. I celebrate that magic in my photography and watercolors."

If the spot where you want to grow a garden is challenging, follow Debra's advice.

Get to know your microclimates. Before you buy plants, notice your lot's orientation to the sun in every season. Patches that get hot noon sun in summer might be shaded in winter. Or if you have a slope, low points may turn colder than spots higher up. Watch how rain falls and where it collects or runs away, and make sure that any slope won't erode during heavy downpours. Learn what your soil is like—whether it's rocky, sandy, boggy, or heavy clay.

Sculpt the terrain. Debra mounds her soil into berms and valleys. With mounds, you can display plants more dramatically, and the added height provides better drainage— especially good for heavy clay soils, like Debra's. To make mounds, bring in several yards of clean topsoil, amend with pumice, and pile it on top of a patch of unpromising dirt. Once you've finished planting, top dress your bare soil with crushed rock or gravel.

Choose the right plants for your location. You can research plants from around the world that grow in conditions similar to your own and try some of them. Or, if you love succulents like Debra does but you live in a colder climate, shop for varieties like hens and chicks *Sempervivum* Chick Charms, which are hardy to Zone 4. In humid climates like Florida, try *Aloe arborescens*, jade plant (*Crassula ovata*), *Furcraea foetida*, or *Kalanchoe blossfeldiana*.

When you find what works, buy more. Plants that grow well in response to the natural conditions where you live tend to stay healthier without much fuss. So if certain plants thrive where you live, choose related varieties. For instance, if you find that mophead hydrangeas (*Hydrangea macrophylla*) are easy to grow, try planting oakleaf hydrangeas (*H. quercifolia*). Debra took on her challenging location by experimenting—and failing—until she found not only what worked, but what she loved.

Debra Lee Baldwin is an author, artist, and succulents expert. To learn more about her plant passion and get plant advice, visit her website, debraleebaldwin.com.

Choosing Plants for Healthy Gardens

When deciding what to plant, take your design cues from the natural landscape. In all but the most arid regions, bare soil gets covered rapidly, whether by wind-blown seeds or sprawling opportunists. In forests dominated by trees, plants lower down have evolved to deal with scarce sunlight. Open meadows and grasslands have other plants weaving around their feet, working together as a community. Even in deserts, where plants space themselves far apart in order to utilize scant resources, all niches, from tall to short, are represented. A layered garden requires far less maintenance.

Check out any layers—or *plant ladders*, as some experts call these sky-to-ground arrangements—that might already exist in your borders or garden periphery. The two plant layers most commonly missed? The first is trees that are in scale with buildings. The second is groundcovers, humble plants that are often not as showy as plants on other layers, but as bottom-rung low-growers, they form a living mulch that mimics the cover found in natural areas. Even containers can be layered: Tuck the largest plants in the pot's center and smaller ones toward the rim.

When you plan for layers, and you also build in natural resilience through diversity. You can choose plants at all levels with berries, flowers, or foliage to attract pollinators and other fauna to your garden. By considering all the rungs on the ladder, your garden gives the strongest support for the wildlife in your area.

Build in Biodiversity

Plants that have already evolved in your region are generally easier to tend. The insects and birds that inhabit your garden have most likely evolved with these native plants. Even the microbiota in the soil have worked together over eons to create a vital interaction between plants and those that depend on them. Any time you choose a native, you support and strengthen this ecosystem.

Still, a garden of all natives can have drawbacks. One is seasonality. For instance, in the parts of the country with little rainfall in summer, the growing strategy of most native plants is to

Opposite: **Plants make the hillside hike fun to explore.** Succulents in shades of green flank these stone steps leading up a steep hillside in Debra Lee Baldwin's garden (page 44). On the left side, *Crassula multicava* plants bloom above various aeoniums. On the right side, big rosettes of yellow and green *Aeonium* 'Sunburst' accent the smaller rosettes of *Aeonium* 'Kiwi' and the green *Aeonium haworthii*.

Overleaf: **Grow a rock garden.** Where soil is stony and dry, consider building and planting a rock garden, as David Salman has done in this New Mexico garden. Start by laying large rocks in a circle about four feet across. Then add smaller rocks on top, filling in around them with soil, and then plants. The plants here include rock wormwood (*Artemisia rupestris*, right), red-flowered *Penstemon pinifolius*, and Turkish stonecress (*Aethionema schistosum*).

appear at the time when rain occurs, and then disappear or stop putting out flowers or growth in the drier months. That can leave gardeners with a reduced palette just at the time of year when they want to be enjoying the outdoors.

Another drawback is availability. Garden-worthy native plants are sometimes hard to find, so support your local native plant sources and spread the word when you find them. Join local native plant groups online. Many knowledgeable participants will generously share seeds and starts. Visit native plant nurseries and you will be repaid with plants that will bring you closer to the land, even if you're only growing them in three containers on your back steps.

But beware—not all native plants are suitable. Some resist "taming" because their needs cannot be duplicated in a garden setting. For instance, the tiny Northwest native fairy slipper orchid (*Calypso bulbosa*) grows under Douglas fir trees, in a soil that perfectly supplies its needs. That miniscule orchid turns up its toes if taken out of the wild and transplanted to a garden.

Neil Diboll, president of Prairie Nursery, the biggest supplier of native plants in the United States, says, "We try to avoid selling the 'specialist' plants because gardeners will be disappointed." Other natives are too rambunctious to be allowed. "These include rhizomatous species that spread by underground runners, often at the expense of other, less aggressive species," Neil says, "and a few species that self-sow seed and can take over a planting." However, that leaves plenty of well-behaved natives that can be great garden companions. Many of them will blend in well in home gardens.

Of course, most of the plants we see in garden centers and offered online are cultivars, cultivated varieties of plants bred to solve problems. Plant breeders have a long and illustrious history of producing plants with more vigor, better color, more flowers, more fruit, or longer flowering times. Sometimes the results are a mixed bag, such as breeding tomatoes that travel well but have no flavor. Sometimes, plants just get away from their breeders. Luther Burbank's Himalayan blackberry (actually from Armenia) is an example; it runs rampant in the Pacific Northwest, where it's considered a noxious weed. In fact, it's so tough it has been used as a center barrier on the freeway.

But what if you want both natives and non-natives in your garden? A six-year study conducted by the University of York, the UK Centre for Ecology and Hydrology, and the Royal Horticultural Society has shown that a mix of natives and non-natives in a garden actually provides stronger support for wildlife, specifically pollinators. There's no reason to think the same wouldn't be true in the United States.

For instance, after a native penstemon (*Penstemon digitalis* or *P. serrulatus*) has flowered, a nearby hybrid penstemon such as *P.* 'Red Riding Hood' or *P.* 'Dark Towers' can not only brighten your border for an extended period of time, but also give pollinators a wider array of nutrient sources to prolong their season. Plants that originated in Europe support our honey bees, which are also not native to the US. The study concludes that "a balance of both native and non-native plants may help provide a home for the widest variety of insects in our gardens."

It turns out the healthiest gardens mix natives with those that have evolved in other parts of the world, and those that have been bred to catch your eye at the garden center. You can welcome them all, knowing that you're supporting the widest diversity of plants for your locale. And that makes your bit of earth a partner in the revitalization of the health of all.

Healthy gardens start with robust plants. At the nursery, buy only certified disease-free plants with lush foliage and well-developed root systems. Make sure that the plants you want to purchase are not are pot-bound, with roots poking through the drain holes, and that they aren't harboring aphids or other pests.

By dealing with an independent garden center you'll find knowledgeable employees who can pick out excellent plants for you. And if you return a plant that isn't up to snuff immediately—not after months languishing in its original container—most reputable nurseries will exchange it.

If you're online shopping, look for mail order nurseries that have been in business for years. The range of online offerings is so great that you need to read plant descriptions carefully, to make sure the tree, shrub, or perennial you want is suitable for the space you have. Keep in mind there's a limit to the size of plants that can be shipped.

Online or in the nursery, if a plant's label says it's full sun—six to eight hours a day—but you need something for a shaded bed, keep looking. Note what the plants need, not only light, but also location (wet or dry) and soil (clay or sand). There are plants that will grow in all these places. Reading descriptions gets you started; then follow up with reliable websites or plant encyclopedias.

Focus on natural materials. Where it makes sense, draw inspiration from the natural materials around your garden before you bring in furnishings. In this Washington State garden, tall conifers set a woodsy tone for the garden's nature-inspired design, so boulders and smaller stones make perfect accents. The stones all look as though they were set there by nature, as do the low, mounding shrubs around them. Stone viewing seats accent the scene and make great recliners. Purchased at a local antique shop, they are ancient stone "cradles" from China, where they once held the poles on which to wind oriental carpets.

Nancy Buley

CHOOSE THE RIGHT TREE— ESPECIALLY WHEN YOUR SPACE IS SMALL

"The best trees are ones that that provide all-season appeal, such as spring flowers, fall color, fruit for birds, and then beautiful winter bark." So says Nancy Buley, who sees rows and rows of those trees daily in her role as communications director for J. Frank Schmidt & Son Co., a premier tree grower in Boring, Oregon. Plant a tree even if you have a small garden, she says; selections for small trees have never been better. The best part: Newer introductions of trees are healthier; most require no spraying to keep bugs or disease in check, thanks to breakthroughs in plant breeding. Nancy says, "Messy fruit is not always an issue anymore, either."

After growing up among trees near Mt. Hood, Oregon, Nancy became more fascinated with them during walks as a student at Oregon State University. Now a powerful advocate for trees, she and the volunteers from her Friends of Trees organization have planted more than a million street trees over thirty years in Oregon neighborhoods. "It's peaceful to live among trees," she says. "Trees have a calming effect; just being around them makes you feel good."

So what do you plant if you have a small yard? Look for trees that top out at fifteen to twenty-five feet tall. Always keep in mind that "trees are like kids," Nancy says. "You turn around, and they're all grown up."

Select the right tree for your site. "There's no such thing as a one-size-fits-all tree," Nancy says. Before you buy, consider your climate zone, rainfall, soil pH, the tree's growth habit (whether columnar or spreading), and the water quality in your area. Then consider your personal preferences for flower, fruit, fall color, texture, and other attributes you'd like your tree to have.

Seek expert advice. A tree is an investment in both time and money. Ask for recommendations from your local independent nursery, or get guidance from local or regional botanical gardens; many have tree selection guides. Cooperative extension services offer excellent gardening information on their websites or from their trained volunteers—Master Gardeners. Every state has an Urban & Community Forestry Department, supported by the USDA Forest Service. Another good source: The International Society of Arboriculture.

Consider new varieties. New upright crabapple trees grow eighteen feet tall and only seven feet wide: *Malus* 'Raspberry Spear' has purple leaves and flowers, and *Malus* 'Ivory Spear' has green leaves and white blooms. Dwarf crabapple *Malus* 'Sparkling Sprite' makes a perfect little topiary when young and is tidy and formal, growing to twelve feet high and wide. 'June Snow' dogwood (*Cornus controversa* 'June Snow'), a J. Frank Schmidt introduction, has strong horizontal branching and sends out clusters of white flowers in June, followed by blue-black fruit and flame-hued leaves in fall. "That tree is a personal favorite of mine," Nancy says.

Grow elegance. A newer fifteen-foot Japanese snowbell (*Styrax japonicus* 'Evening Light') has unusual dark purple-green to maroon foliage with white flowers, followed by decorative nutlets that give a spark to autumn bouquets. "This one takes humidity and heat," Nancy says.

Or choose an old favorite. Upright-growing Japanese maples are lovely small shade trees, with a graceful form, and great foliage and fall color; they thrive in filtered shade. *Acer* 'Seiryu' is an upright form to twelve feet. In colder climates, try Korean maple (*A. pseudosieboldianum*), Northern Glow, or a paperbark maple (*A. griseum*), a fifteen-foot handsome small shade tree with peeling orange bark.

Buy your tree in a container. If you are new to planting trees, buy container-grown rather than bare-root, Nancy advises. They're easier to get established and aren't as dependent on even soil moisture through the first growing season. But make sure the plant has been grown by a reputable nursery that doesn't leave trees in containers for too long, a practice that leads to circling roots, which can cause the tree to die years later. Container trees can be planted almost any time of year as long as they are adequately watered and cared for.

In addition to her day job at the J. Frank Schmidt nursery, Nancy Buley is co-owner of Treephoria, a smaller tree nursery (treephoria.com). Find more about the benefit of trees at www.jfschmidt.com.

Made for the Shade

If you're new to gardening, it may seem that all plants you see for sale are those that grow in full sun. Yet so many of us have shady areas—the overhang, a porch, or in the shade of a tree. These are the spots where you would like to relax on hot days, as well as the places where many terrific plants like to grow. Finding them is just a matter of looking beyond the first dazzling plant that catches your eye.

Many small trees have evolved to occupy the shaded understory niche; these include American plum (*Prunus americana*), with its white spring flowers, and white fringe tree (*Chionanthus virginicus*), which grows fifteen to thirty feet and puts out wispy but highly fragrant flowers. Shade is also the place for drama, with large-leafed plants like hosta, spotted lungwort (*Pulmonaria* spp.), epimedium, with their tough leaves and delicate fairy flowers, and long-blooming hellebores, whose large palmate leaves can carry though the seasons.

Contrast those with the fine-needled look of conifers that will tolerate shade, such as a Japanese umbrella pine (*Sciadopitys verticillata* 'Joe Kozey') or a low-growing yew, Emerald Spreader (*Taxus cuspidata* 'Monloo'). Fill in with plants like spreading bunchberry (*Cornus canadensis*) or the dark *Ajuga* 'Chocolate Chip'. And don't forget the wide choices for ferns, from six-footers down to tiny groundcovers, and you've got a shade garden with year-round good looks.

So don't let shade conditions be daunting. If you've got shade, go on the hunt for those plants that will love to settle down there, and you can join them, with a glass of something cool and a good book.

For Easy Care, Find the Right Shrubs

From shrubs that could double as small trees down to little two-footers, these are the plants that will fit into any spot where you wish to garden. Once they're established, if placed appropriately—not too big for the spot, and in the growing conditions they need—shrubs are about as carefree as any living thing can be. You might cut off spent blooms, but continual deadheading will usually not be involved. You won't have to divide them every few years.

Layer your plants. For easy viewing, arrange plants in beds and borders by height, with the lowest ones in front, mid-size shrubs behind them, and the tallest shrubs or trees in back. In this leafy corner of a Pacific Northwest garden, foliage plants in soothing shades of green fringe the path; they include clumps of chartreuse Japanese forest grass (*Hakonechloa macra*) and hosta, whose large clumps of big, bluish leaves make them nearly perfect understory plants, along with ferns and shade-loving perennials.

Many shrubs are tough. There's a reason lilacs and forsythias are found growing happily near abandoned farmsteads. Shrubs will be there year after year with minimal feeding and fussing, giving you more time to enjoy your outdoor space.

When it comes to buying shrubs, the displays of them in garden centers are often not as alluring as those for annuals and perennials. But they will be with you for a long time, so investigate shrubs with the same care as you would trees. Plant breeders know you are looking for small size and ease of care, and they have responded, bringing many varieties to market in the past few years. Shrubs that bloom more than once are especially popular in azaleas, lilacs, and hydrangeas. Here are just a few of many choices you have.

Hydrangea: These beloved shrubs are known for their showy blooms in sun to part shade, depending on in what part of the country they're planted. Most need regular water; their name even has *hydro*, the Greek word for water, in it. Breeders have been hard at it and now there are breakthroughs in size, like the compact three-foot *H.* Bobo, or the two- to three-foot rebloomer, *H.* Let's Dance Big Easy—both great for containers.

Roses: Roses are arguably the most beloved of garden plants, and most people have favorites among the hybrid teas or antique selections. However, there's a whole new world out there of tough, disease-resistant roses that require no special attention to look good and reward you with repeat blooms throughout the season. These new specially bred roses are often called *landscape* or *easy-care* roses. Ping Lim, an award-winning rose hybridizer with his own line of both tough and highly scented roses, says, "Black spot is our friend." How can a noxious disease like black spot work *for* him? In his breeding program, any potential rose that displays signs of this disease is thrown out. Ping's Easy Elegance line and his newest True Bloom selections feature the best that roses can offer.

Weigela: They're showy, long-blooming, and long-lived. These shrubs are deciduous, although they can hold their leaves in parts of the country with warmer winters. Some have colorful foliage as well as flowers. The trumpet shaped flowers are loved by hummingbirds. Plant in full sun to light shade. Tall ones, like *Weigela florida* 'Briant Rubidor' extend to seven feet, while many, like *W.f.* 'Alexandra' Wine and Roses, with its dark maroon leaves and pink flowers, stand about five feet tall and wide. Even smaller, with the same dark leaves, *W.f.* 'Elvera' Midnight Wine tops out at twelve inches by eighteen inches—perfect for the front of the border or containers.

Marietta and Ernie O'Byrne

PAINTING WITH PLANTS—CELEBRATE PLANT DIVERSITY

A healthy garden features a diversity of plants—natives and non-natives, deciduous and evergreen, all shapes and sizes mingling together—from trees down to groundcovers. Marietta and Ernie O'Byrne, owners of Northwest Garden Nursery, have filled the beds and borders around their 1918 farmhouse with an astonishing array of plants. There's something of interest in their garden year round, whether it's blooms, berries, nuts, or grasses. That diversity attracts a multitude of birds and pollinators, which this intensely plant-loving couple enjoy—right down to the little green frogs in their water storage tank.

Clearly, Marietta and Ernie have elevated their garden to artistic heights by combining plants for color, shape, and texture, with striking results. It's as if they're using plants as an artist might use paint on a canvas.

You can plan for this kind of artistic diversity too. The same principles apply, whether you're arranging a container of mixed plants or a whole garden bed. Here are Ernie and Marietta's tips.

Combine plants that share similarities. All plants that you want to group in a particular spot need the same conditions for light, soil, and water. But the artistry happens when you also group plants that have similar structures, or blend together plants with similar leaf colors, or fill a bed with flowers in the same color range. Together, these strategies create a sense of harmony in the plantings.

Contrast foliage shapes, sizes, colors. For drama, set plants with different leaf sizes and shapes beside one another—a large bold leaf next to a tiny one, for example. Choose flowers whose colors and shades play off one another, such as vivid yellows next to intense blues. Marietta admits that she sometimes wakes in the night, mulling over new plant combinations. She asks, "How about a blue flowered *Vitex agnus-castus*, coupled with golden *Rhus typhina* Tiger Eyes plus the dark blue *Agapanthus* 'Storm Cloud'?" Now that's beautiful drama.

Note plant combos you love. Marietta and Ernie are always on the lookout for new plants. And they're constantly figuring out new ways to combine them. You can too. Keep a wish list, either online or in a notebook. Join social media groups where others show pictures and talk about plants. When you find one that catches your eye, note its best qualities—flower or foliage color or singular shape—and add it to your list. Then look for others that show both similarities and differences. You can start to visualize groupings before you buy.

Don't be afraid to experiment. If Marietta and Ernie's combinations don't pan out, they simply move plants to another location—although, they caution, not during summer heat. "Be willing to lose some things," Ernie advises. Marietta and Ernie's garden is living, growing, and changing, in the most elegant and beautiful way—and yours can too.

Marietta and Ernie O'Byrne are the owners of the wholesale Northwest Garden Nursery and known for their line of Winter Jewels hellebores. Their book A Tapestry Garden—the Art of Weaving Plants and Place *describes their garden adventures.*

Opposite: **Blend shades of one color.**
A cheerful red metal chair sets the autumn tone for this collection of plants in the O'Byrne garden. Behind it, a tall iron column lifts a potted hair sedge (*Carex flagellifera*) skyward, where its orange foliage echoes the turning fall color of the dawn redwood at right (*Metasequoia glyptostroboides* 'Gold Rush'). Adding shades of the same hues are blooms of pink *Daphne × transatlantica* 'Eternal Fragrance' on the left, and the reddish stems of *Helleborus × sternii* in the foreground.

Above: **Use similar colors with contrasting leaf shapes.** The starburst-shaped leaflets of a silver-leaf Himalayan cobra lily (*Arisaema consanguineum*) partners with rounded, frosty-leafed *Brunnera* 'Looking Glass' below.

Plant for bees. Grow flowers that they love, and bees can become your garden's best allies. As they buzz about and alight on blossoms to sip nectar, they gather pollen that they spread to other blossoms, pollinating your plants for free. In this garden, foxgloves, larkspur, and other bee favorites surround a hive that's nestled beneath a tall, pink-flowered rhododendron, making foraging easy. A second hive is nearby. To extend the pollinator season, plant a succession of bee-friendly blooms—say, lavender and rosemary for spring; followed in summer by coreopsis, penstemons, and purple coneflower (*Echinacea purpurea*), with *Aster × frikartii* for fall blooms.

Conifers Have Your Garden's Back

In many climates, evergreen cone-bearing plants are like good friends—reliable, hardworking, and always around when you need them. Conifers offer great structural support, especially in colder areas where the rest of your garden has gone to ground in winter. Their tidy evergreen shapes bring order to planting exuberance.

Have fun with foliage. In many climates, the easiest borders to tend are made up of beautiful shrubs, grasses, and trees instead of flowers. But keeping all that foliage from disappearing into a green backdrop takes artistry. For the border pictured here, in Edmonds, Washington, garden designer Stacie Crooks selected plants whose shapes, sizes, and leaf colors and textures contrast with one another. They include (clockwise from left): cardoon (*Cynara cardunculus*); brown-leafed smoke bush (*Cotinus* 'Grace'); Japanese gold cedar (*Cryptomeria japonica*), *Sedum spectabile* 'Autumn Joy'; Japanese forest grass (*Hakonechloa macra*); and nandina (*Nandina domestica* 'Moon Bay'). A Chinese windmill palm and an American chestnut tree grow in the background.

Upright, or fastigiate, forms are especially important for structure. They break up long horizontal sweeps of plantings with sky-piercing challenges. Tall, thin conifers frame some views and block others. When you don't want to be distracted by anything beyond your fence, a narrow white pine (*Pinus strobus* 'Fastigata') will keep the attention on its blue-green foliage inside your yard. The naturally occurring geometric forms of cones, rounds, or columns work well in formal situations that rely on plants of similar shapes to mirror each other.

In addition, many conifers grow in mountainous regions of the world. These alpine specimens will be contented in the drier confines of large containers. Their slow-growing habit allows them to be there for years before they need to be planted in the garden. But remember, all conifers always keep growing—whether in inches a year for dwarf varieties, or in feet for the taller ones.

Add Grasses for Motion and Beauty

Moving in the slightest breeze, ornamental grasses are another piece in the healthy layering of plants. Their fine texture contrasts well with thicker-leafed plants, and many are upright, their fluttering seed heads held high. Most are easy-care; one cutting down at the end of winter if you've left them standing, and the show starts again.

When hunting for good grasses, especially online, check for invasiveness. For instance, Mexican feather grass (*Stipa tenuissima*) is a native in some areas, but it's a noxious weed on the invasive species list in California and other parts of the west due to its overwhelming number of seeds. The good news—plenty of other terrific choices can duplicate its airy habit. Try blue grama grass (*Bouteloua gracilis* 'Blond Ambition') with its chartreuse flowers that pass to golden seed heads. Or consider bull grass (*Muhlenbergia emersleyi* 'El Toro'), with the added benefit that it will grow in part shade.

For an upright grass that will squeeze in just about anywhere—even in a two-foot diameter container—the old standby is *Calamagrostis × acutiflora* 'Karl Foerster'. Another contender for an upright accent is Northwind switch grass (*Panicum virgatum* 'Northwind') with green stems that grow six feet tall and only three feet wide and turn golden in the fall. With the easy care popularity of grasses, there's sure to be one—or many—just right for your gardening situation.

Janet Sluis

THE SECRET TO SHOWY BORDERS—BLEND PLANTS FOR COLOR, TEXTURE, AND MOTION

Janet Sluis puts together plant collections the way a fashionista might choose accessories and shoes to go with a dress: carefully, but with pizzazz. That's why the perennials, shrubs, and grasses that she selects for eventual sale at retail nurseries are so colorful, richly textured, and shapely.

Unlike fashions that come and go, her plants are not only stylish in beds and borders; they are also tough and long-lasting.

But, then, Sluis is no stranger to the nursery business. Growing flowers and vegetables could well be in her genes. Her father's family operated a seed business in the Netherlands. And her father, after immigrating to America and then settling on a five-acre property in San Bernardino, California, became a grower of ornamentals too.

As curator of plant collections for Plant Development Services Inc., she uses her plant pairing skills to select gorgeous shrubs, perennials, and grasses that are sold in curated collections at retail nurseries. She has perfected the art of combining these plants in beds and borders with dazzling results. Blue grasses beside golden-leafed shrubs, for example, perhaps edged with a splash of lime-hued grasses. And borders filled with her favorite hues—shades of blues, purple, and soft green.

Whether you are planning to plant a small square bed or a big border, consider some of her strategies for choosing and pairing plants.

Analyze your space. Before you head off to the nursery or garden center, Janet suggests checking out the patch of ground you'd like to fill with plants. Is it just the right size for one plant? Or maybe four in a small square bed, perhaps to view from all sides? Or is it a long, narrow border, perhaps with its back against a fence? Does it get full sun most of the day, or part shade, or full shade? Measure the length and width of the space you have in mind for your bed or border.

No space? Go portable. "My entire driveway is lined with beautiful plants, all in pots," Janet says. "Container gardens offer a great solution where soils are poor. Or, if you have a hole in your border, try setting a potted plant there."

Choose the right plants for the right place. Before you purchase, read the labels. Make sure that the ones you choose will thrive in the spot you have in mind. Also consider how quickly, and how large, each plant will grow.

Pick your color palette. At your nursery, stroll among the plants. Check out the perennials, grasses, and small shrubs. Pick up the one whose color catches your eye. Keep in mind that some colors are harder to work with—reds and pinks, for example. And limit your colors; too much in a garden can look overwhelming. Remember that green is a color too. In very hot climates, gray-leafed shrubs and perennials are generally tougher and less thirsty than others. Once you've found your favorite plant, Janet says, put it in your nursery wagon, then set other plants that might go with it right in the cart to see how they'll look together. Make sure all have similar needs for soil, water, and exposure to sun or shade. If you are ordering online from a nursery, you can still create combinations virtually, collecting photos of companion plants for your cart.

Janet Sluis is the curator of plant collections for Plant Development Services Inc. Her choices are found in retail nurseries around the US.

Cool hues set the theme for this pair of all-season borders in Sonoma, California. Blue-flowered veronica mingles with seaside daisies in the foreground. Other stars of the show include: blue salvias, spiky Platinum Beauty lomandra, and clumps of variegated 'Meerlo' lavender joined with grassy blue fescue. Dwarf rosemary edges the borders in back.

Use Ferns to Bring It All Together

Granted, ferns are not usually the first thing on a gardener's wish list. Yet their richly textured fronds enhance gardens the same way their presence enhances florists' bouquets—they unite disparate plants, bringing them all together harmoniously.

Ferns grow in all situations, from hot deserts to cool shaded areas. Try planting three-to-five-foot ferns such as *Dryopteris × complexa* 'Robust' in a sunny spot, among a group of colorful perennials. You can support lax growers like delphiniums with the sturdy fronds. The ferns will last longer through the growing season than what they're holding up.

Audition Your Groundcovers

Groundcovers put the final finish on a healthy garden. Sometimes called "living mulch," they fill in that bottom niche at the fronts of borders and edge paving beautifully. Their shallower roots don't compete with the larger plants around them, and they hold moisture and soil, preventing rain from pounding and compacting the earth. The best of them block out weeds.

When you cover all the spaces in your garden with something growing, you become more like a referee than a weed puller. You can gently push back those groundcovers that want to lean on taller stems. You will spot the occasional unwanted weeds before they can get established. And the groundcovers discourage germination by keeping seeds in darkness. To get ground-covers established, you will need to mulch the spaces in between the plants. But as these low-growers spread out, you will find yourself adding less mulch, and, in turn, weeding less.

What makes a good groundcover? Height, or lack of it. Most good groundcovers stay under a foot tall. The best ones tolerate varying light conditions. They take sun early in the season when other plants nearby are just starting to grow. Later, they tolerate shade as neighboring plants reach full height. Good groundcovers are well-mannered. Those that are too vigorous can crowd out neighboring plants. Those that are too timid won't fulfill their job of blocking weeds.

Provide for wildlife. This garden holds everything that birds and other beneficials need—shelter, water, and a diversity of food sources. Birds preen in the recirculating stream's shallow spots, and butterflies puddle at the soft shoreline. Protection comes from the outspreading branches of blue lacecap hydrangea at the top of the stream, and food includes the seed heads on the paper reed papyrus (*Cyperus papyrus*) standing in the pool. In a small yard, you could replicate this with a simple birdbath surrounded by short flowering shrubs and a hanging mixed seed dish.

When you first buy groundcovers, you may want to experiment to see which ones will work for you. Buy a few, try them out. Buy more of those that work best. Don't buy plants that have descriptions like "spreads indeterminately." Even slow-growing plants may get out of hand, depending on your situation. If you see too much rampant growth, immediately tear out the few you've purchased—no harm done to the rest of your garden.

Many online sources will tell you which unruly plants to avoid in your location, but often it depends on what other plants share the bed. For example, sweet woodruff (*Gallium odorata*) is a great groundcover for naturalistic shade gardens, especially when paired with large woody shrubs like hydrangeas, big ferns, or even hellebores, which can punch their big spreading leaves right through it. But woodruff's fast-creeping roots can tromp all over less vigorous perennials. In general, the smaller or more delicate the other plants are in a bed, the smaller the foliage should be on the groundcover. The fine leaves of *Herniaria glabra* 'Green Carpet' or even finer Scotch moss (*Sagina subulata* 'Aurea') would let more delicate plants come right through without competing.

The hunt for groundcovers can take you down all sorts of rewarding byways, but when you find those that play well in your garden, you'll be creating a healthy plant community that will make your gardening life that much easier.

Choose a Mixed Lawn

Whatever size lawn you have, you can break up the monoculture if you encourage a mix of grass and small flowers, such as tiny daisies (*Erigeron* spp.), microclover (*Trifolium repens* var. 'Pirouette' or 'Pipolina'), or dwarf chamomile (*Chamaemelum nobile* 'Treneague'), by sprinkling seeds or tucking plugs into your existing grass. Without tearing out what you have, you can also over seed your lawn with one of the newer low-maintenance grass seed mixes that require less water. These vigorous newcomers can gradually outcompete the older grasses by putting down longer roots. That will make your lawn sturdier, healthier, and less water reliant.

Bury Bulb Treasures

Bulbs are available for all light conditions—from sun to shade. There are anatomical differences between bulbs, tubers, corms, and rhizomes, but any plant with a swollen root system for nutrient storage and reproduction is commonly called a bulb. The variety of bulbs is so great that almost anything you put in the ground will be rewarding. For a succession of spring blooms, choose from early, middle, and late categories. Also choose bulbs of various sizes that can be layered in the soil, with the biggest deeper down, and smaller near the surface. For instance, dig in large narcissus on the bottom, then plant tulips, and above that, a mix of smaller bulbs like spring crocus, scilla, and snowdrops.

Here's an easy method for incorporating spring bulbs among existing perennials, grasses, and deciduous shrubs. Think of those plants as islands. The bulbs will encircle them like water in wave patterns. By the time the early show has finished, the burgeoning leaves on the other plants will obscure flopping foliage.

In spring, golden trumpets of daffodils, purple drifts of grape hyacinth, blasts of red tulips, and elegant sheets of scented white snowdrops will splash over your winter-grayed garden. At later times in the year, summer bulbs like lilies, dahlias, cannas, and callas will make their appearance. Even if the tender ones must be lifted and overwintered in a frost-free place, the beauty and great variety of bulbs make them worth it every year.

Melinda Myers

PLANTS WITH BENEFITS—CHOICES FOR POWER-PACKED PERENNIALS

Lively and engaging, Melinda Myers has been talking about plants for more than thirty years as a national gardening television and radio host. She is also the author of twenty books. She gardens in Wisconsin, and says, "I'm always looking for plants with multiple benefits." Perennials that will return year after year are an excellent place to start.

To find what you'd like, Melinda suggests checking out the Perennial Plant Association's choice for Perennial Plant of the Year. "These are selected," Melinda says,

"for their adaptability to various regions, their outstanding aesthetics, and their performance in the garden." To find the list, visit perennialplant.org/page/PastPPOY.

Here's Melinda's sampler for perennials with multiple seasons of interest and appeal for pollinators and birds—and, of course, gardeners.

Coralbells (*Heuchera* spp.): "Even small-space gardeners can start a collection of these plants," Melinda says. "Those in hot, humid areas should select cultivars of *Heuchera villosa* that are more heat tolerant."

False blue indigo (*Baptisia australis*): It has blue-green pea type foliage, blue flowers, and black seed pods. "In winter, the seeds rattle when the wind blows," Melinda reports. "It sounds like wooden wind chimes."

Milkweed (*Asclepias* spp.): The only food for monarch caterpillars; bees, butterflies, and hummingbirds also like them. Melinda says, "*Asclepias syriaca* is an aggressive grower, spreading by seeds and underground rhizomes. *A. sullivantii* is a naturally occurring variety that is less aggressive." Swamp or red milkweed (*A. incarnata*) likes it moist, and tolerates dry soils with fewer

Melinda favors wild quinine (*Parthenium integrifolium*). In the Lurie Garden, in downtown Chicago, the three-to-four-foot stems put on a long three-month show of fuzzy white flowers that attract pollinators and the caterpillars of both moths and butterflies. Native to the eastern US and the Midwest, wild quinine happily grows in dry to medium-moist soils. Winter bonus: The upright stalks turn deep brown—more seasonal interest for this rare and unusual plant. "It's like a yarrow," Melinda says, "but bigger, and it doesn't take over. It's my new favorite."

tendencies to take over. Commonly called butterfly weed, the orange blooms of *A. tuberosa* are among the showiest of native flowers.

Pale purple coneflower (*Echinacea pallida*): Tough and adaptable (ranges from Zones 3 to 10), bees, butterflies, and hummingbirds are attracted to the flowers, while birds love the seeds. Melinda recommends "Echinacea hybrid 'Cheyenne Spirit', with fragrant flowers in a variety of colors, and Sombrero Salsa Red, which is compact and hardy."

Peonies (*Paeonia* spp.): "New foliage emerges in spring with a tinge of red," she says. The flowers are often fragrant and excellent for cutting, and the large healthy foliage makes a great contrast to other plants' finer leaves all summer long in Zones 3 to 8. "In warmer climates," Melinda says, "choose those that bloom early to avoid summer heat."

Prairie dropseed (*Sporobolus heterolepis*): This easy-care native grass forms a two-to-four-foot mound with late-summer flowers held above. The panicles have a distinct fragrance—some say it's coriander or butter—but to Melinda, "It smells like burnt popcorn, which I like." Plant in dry, hot places; it's attractive to birds and unattractive to deer. "The panicles look like gems when covered in ice," Melinda notes. "We northern gardeners are always looking for any subtle beauty in the winter garden."

Rattlesnake master (*Eryngium yuccifolium*): This perennial has blue-green yucca-like foliage and long-blooming bristly round flowers. "It has great texture and strong form, but still blends nicely in the garden," Melinda says.

Switch grass (*Panicum virgatum*): This bunchgrass has upright summer foliage and late flowering stalks that hold up through winter. "Consider cultivars like 'Shenandoah' (4 to 5 feet) and 'Northwind' (5 to 6 feet) that are not overly aggressive," she says. It's also a host plant for the caterpillars of several moths and butterflies.

Threadleaf tickseed (*Coreopsis verticillata*): A native perennial hardy in Zones 3 to 9, the daisy-like flowers bloom a long time without deadheading. Look for cultivars such as the eighteen-inch-tall 'Zagreb', which Melinda likes. "I find it overwinters better than other cultivars in heavy clay soil."

Melinda Myers is a TV and radio host, speaker, author, and professor on the Great Courses series "How to Grow Anything." Find out about everything she does at her website, melindamyers.com.

Create Colorful Containers

Just about any plant will grow in a container—for at least a little while. Eventual plant size matters, but even smaller roses and slow growing conifers can thrive in a pot for a few years, although the latter won't tolerate being crowded by other plants. Whether you're choosing plants that will give you plenty of enjoyment throughout the year or changing pots out seasonally, container gardening can make an important contribution, no matter how small, to the health of our ecosystem.

You can grow a single plant in a container. Or you can create a mixed arrangement in one large pot that combines a focal point, usually an upright plant, with others that fill in, and some to soften the edges, often called *thrillers*, *fillers*, and *spillers*. Container designers often rely on the classic size proportions of one-third container to two-thirds plant material, but these proportions are only a guide and can be adjusted in many ways for satisfying results.

For containers in sunny spots you might like upright *Canna* 'Cleopatra', with its bright orange flowers, or sculptural 'Golden Sword' yucca. Or pair them in large pots with the paddle-shaped leaves of *Bergenia cordifolia* and the fine white baby's breath flowers of *Euphorbia* 'Diamond Frost'. Tiny white bacopa flowers could cascade over the side as summer goes on.

In shade, a single upright plum yew (*Cephalotaxus harringtonia* 'Fastigiata') can command a spot by an entrance without taking up too much space. Raise up the massive leaves of *Hosta* 'Gentle Giant' to new heights by growing it in an expansive container. Colorful coral bells like deep maroon *Heuchera* Dolce 'Cherry Truffles' can accentuate a pot all by itself, or pair it with silvery *Brunnera* foliage and allow a bellflower (*Campanula punctata* 'Cherry Bells') to spill over.

Grow a window box.

Where space is limited, as in Charleston, South Carolina's Historic District, boxes filled with flowers hang beneath many windows. Each one is a work of art, viewed by passersby on sidewalks below. This all-white combo, for example—it's a single-hued palette made more interesting by varying bloom sizes, shapes, and habits. If you have a place in mind for a window box (and you can easily keep it planted and tended), choose a box made of moisture-resistant material, such as cedar or redwood. Make sure it's at least six inches deep, with holes in the bottom for drainage.

Opposite: **Go easy; plant containers.** What's not to love about container gardens? They're easy to plant and tend, and they bring the look of a garden onto decks and patios. In this garden in Seattle, Washington, flowers in bright reds and oranges accent the deck. The largest container (front left) features red-flowered begonias and blue-flowered lobelia; a spray of chartreuse creeping Jenny spills over the side.

Make magic when night falls.
When a garden space is small and details are important, get creative, as did the owners of this garden in Portland, Oregon. Working with garden designer Marina Wynton, they framed the deck with dogwoods and clustered a few planted containers on their deck. But the drama comes when night falls as they dine outdoors on summer evenings. That's when potted *Phaelanopsis* orchids, set in pockets in the deck's fences and in the side of the house, glow softly from small lights that hang above. All orchids overwinter indoors.

Fill the gaps. All borders present challenges at one time or another; a plant may overgrow its bounds and need to be moved, or it may appear to struggle. When that happens, first select a plant that takes the same conditions as the remaining plants. (This salvia will feel at home among the other dry-climate plants, including lavender and Echinacea). Gently dig up the failing plant with a pitchfork, taking care not to disturb nearby roots, then dig a planting hole no deeper the new plant's root ball. Set the replacement plant in the hole, water thoroughly, then back fill the hole until the surface is even with the newcomer's original soil line.

Caring for Healthy Gardens

Taking care of your garden, big or small, will be much easier if you choose the right plants. By creating a diverse habitat, learning your location's strengths and limitations, and then choosing suitable plants, you can go a long way to making a garden healthy.

But what about dealing with pest, diseases, and weeds? If you commit to following the least harmful practices for controlling these issues, you can keep plants healthy. Still, you can't always control the weather. A long stretch of drought or massive periods of rain can stress plants and make them more vulnerable to pathogens, like powdery mildew. Cleaning up when you can, tolerating some rough-looking plants when you must—it's all part of the game.

When any of your plants show signs of disease or pest, the first thing to do is identify the culprit. Then look to the least harmful treatments, like hosing off the leaves to dislodge sucking insects, or encouraging habitat for beneficial insects. Hold off on more drastic measures. Before you reach for the heavy-duty chemicals (organic or not) that seem to be needed to maintain the health of certain plants, consider a different approach. Lose them. After all, composting is part of the gardening cycle.

But do not toss seriously diseased plants in your compost pile, which may not get hot enough to kill many pathogens. If your area has a green waste disposal, put the plants there. Commercial composting reaches effective killing temperatures of 160 degrees. If your community doesn't have a green waste service, bag the plants, leave the bags to shrivel in the sun, and then put them in the garbage. Burning used to be a time-honored way of getting rid of plant material, but with the awareness of air quality problems, many areas of the country no longer encourage it.

The world is full of terrific, hardy, disease-resistant plants. Give up on those that struggle. You can find plants that will live and thrive where you are without an arsenal of chemicals and a lot of extra fuss.

Healthy gardens grow from healthy soils. Excellent soils contain plenty of organic matter, nutrients, and micro-organisms—the good fungi and bacteria. Together, these elements improve the soil's texture. Like a giant sponge, the soil is porous enough to support plant roots, yet also holds enough moisture to allow roots to feed before the nutrients percolate away. Healthy fibrous root systems that go down deep will slow water movement, not letting it run off and gather pollutants that carry into streams, lakes, or coastal waters.

Gathering of Gardeners

Joe Lamp'l

GROWING HEALTHY SOIL—FEED THE SOIL, AND YOU'LL FEED THE PLANTS

"Every plant wants to have a full and productive life, with vigor and resistance to predators and disease," says Joe Lamp'l. If anyone should know what plants need, it's Joe. From his television show, *Growing a Greener World*, to numerous podcasts and online garden courses, Joe is dedicated to getting the word out about organic gardening and sustainability. And one of the most basic secrets of horticultural health lies unseen around the plants' roots.

That's where soil biota lives. Fungi, bacteria, protozoa, and nematodes are some of these miniscule microorganisms; plant roots attract and feed these mini creatures. In turn, soil biota forms elegant and complicated relationships with the plants, transforming the soil's ingredients into exactly what the plants need for healthy growth. These hidden heroes also change soil structure, binding together particulates into larger pieces that retain water better and allow more air to get to the roots.

So, how do you get these tiniest of garden helpers to work for you? Nourish them by

Load up on garden gold. Sustainability is an ongoing process, not an end result. These wheelbarrows are piled with Joe Lamp'l's favorite mulch—shredded leaves. "When you use organic matter," he says, "you give your plants the most hospitable growing situation they can have."

continually adding organic matter to your garden beds. "It's like maintaining a bank account," Joe says. "You have to make deposits. If all you did was withdraw, you'd go broke. It's the same with soil." Here's how he does it.

Dig in amendments. Joe's favorite is compost, but when you're planting, you don't need a lot. "It's really wasteful to dig it more than six inches down, because most of the roots are at the top," he says. Spring and fall, he scratches in an inch or so of compost into the soil surface around permanent plantings and also adds a sprinkle of ground bark and granite dust, which helps replace trace minerals in his soil.

Apply mulch around plants. This soil builder is essential. Joe would never be without it. "Mulch reduces erosion from the rain and wind; it holds moisture," he says. "It's really composting right in place, because eventually it breaks down into what the soil needs." He favors shredded leaves. In fall, Joe collects his neighbors' bagged leaves as well as his own, shreds them with a mulching mower, then stores them on his property in round wire fencing bins. Shredded bark and wood chips are other good choices, but avoid using dyed mulches, which can hide unknown contaminants from recycled wood products. How thickly should you apply mulch around plants? Joe recommends a layer about two inches thick—no more than four. To discourage plant diseases and keep from smothering stems, leave several inches mulch-free around the base of all plants, from trees down to lettuce.

Avoid chemicals. Strong fertilizers, pesticides, and weed killers can wipe out whole populations of those unseen critters underfoot—one great reason to go organic. But even organic products can be broad spectrum, and may cause harm to the balance in the biota. If you can, it's better to foster the great relationships in the soil food web with few chemicals of any kind. Just keep on depositing in the soil bank, Joe says. Good biota work hard to support your plants' health, so you want to work hard to keep them happy.

Joe Lamp'l lives on a five-acre property, GardenFarm, in Georgia. For more of his great advice, visit his websites, GrowingAGreenerWorld.com and JoeGardener.com.

Many garden soils don't meet this ideal, but soil can be improved. It's especially important for heavily-used vegetable beds, where digging in fresh amendments is usually necessary every season. However, keep in mind that any time you dig up the soil, you bring chaos to the unseen web of soil biota and introduce extra oxygen, which breaks down the amendments too quickly. The healthiest soils are those that mimic nature and are left undisturbed. So, for most other ornamental beds and borders, once they're established, don't dig, unless you need a hole for planting. Amendments may be added as mulch on top. Earthworms and other creatures will obligingly mix them in for you. Add stepping stones or pathways right into your beds so you can stand and work without compacting the soil around the plants' roots.

What Happens in Your Garden Stays in Your Garden

Autumn leaves, shrub branches, grass clippings, old flowers, and vegetables—your garden is generating an enormous amount of spent material, all of which still has goodness generated by the sun. And the best place for most of it is right there, where it grew in your garden. The trick is, garden waste needs to decay. Sometimes just tucking the fallen leaves around the plant's base is enough. For instance, when fern fronds die at season's end, they can be left as a protective mulch, and next year's fronds will arise right through, hiding the base. On the other hand, if your roses show diseases like black spot and mildew, a fall cleanup and a good layer of mulch is in order so those pathogens don't overwinter underneath the bushes.

Woody material will need to be chipped into smaller pieces in order to break down in the soil in a timely way. Instead of lugging your tree and shrub clippings out to the curb for yard waste pickup, perhaps you and your neighbors can co-own a chipper, passed from house to house as needed. Put your mulching mower to work shredding leaves by going over a pile, reducing the size. Perfect free mulch. Allow that same mower to feed your lawn as the grass cuttings sift back down to the soil. And vegetable household scraps can also be a component in your garden through composting.

Composting is a method of breaking down plant material into nutrients plants can use. It's about layering. It's that simple. A green (wet) layer like vegetable scraps (avoid protein) or grass clippings is added to a brown (dry) layer like leaves, spent flowers, or old stalks. There are whole books on this process, but basically that's it. Start it by making the first layer—what

Organize your garden tools. Your garden life is better if tools are kept handy and clean. It's easy to wipe off fresh mud with a rag before it dries. And for your garden's health, keep a bottle of disinfectant (alcohol works well) nearby for cleaning shears and loppers when pruning, as you move from one plant to another. That simple practice prevents diseases from spreading.

biodynamic farmers call a "mattress"—out of loose woody stems with plenty of open spaces for air circulation, and then build one wet and one dry layer at a time.

Studies have found that the most productive composting comes from open heaps, but in suburbs and cities, you should choose a container that is rat-proof so you're not creating an attractive nuisance in your neighborhood. Faster results come from turning the ingredients, but you don't have to. Slower composting can retain more nutrients, but it's good to have two containers. One to actively fill, and one to sit and decay. You might share the cost of critter-proof containers with another household.

Care for Containers

All the methods for encouraging healthy plant growth are the same when growing in containers—fulfilling their basic needs for light, food, and water. But the stakes get higher when you're growing out of the ground. Container plants dry out faster, freeze more readily, and run out of nutrients without being able to replenish them in their confined space. On the other hand, containers are moveable, so you can position them in whatever light the plants need, or put them near a reliable water supply—drip tubing and a simple timer on a faucet is an easy-care solution—and tuck them into a frost-free place in harsh winters. For plants left outside, be sure to water if the soil dries out in winter. If you're moving larger containers, a good plant dolly is worth having. In small spaces, you might share one among neighbors.

You can successfully grow plants in all kinds of materials—from fancy Italian terra cotta to heavy-duty soft-sided grow bags. In cold winters, make sure your containers are made of high-fire pottery, concrete, or other frost-proof material, as the water in soil expands when frozen and can crack less-sturdy pots.

And don't forget pot feet. To keep from staining decks and patios, and to allow air circulation under containers on the ground, you need to elevate your pots. Traditional pot feet are made of terra cotta or high-fired glazed clay. Usually three support one round container, sticking out around the base. Modern pot feet are plastic rings or other risers that lift the container without being seen.

Bob Lilly

GARDENS ON WATER—ALL CONTAINERS, FOR EVERY SEASON

Seattle's Lake Union: That's where horticulturist Bob Lilly is at home. He has lived in a houseboat there since the 1970s. Because plants are his passion, this designer extraordinaire (and president of the Hardy Plant Society of Washington) has turned his deck into a garden. Each summer, his shake-shingled dwelling disappears beneath a cascade of greenery and blooms, with all plants grown in containers.

Like many who arrive in the plant world, Bob prefers hands-on knowledge, which is why, when he graduated from college, he went to work at nurseries.

Here's Bob's advice for container success.

Try out potting soil brands until you find the right one. Not all are created equal. Some are too heavy for good root growth, or the ingredients break down too fast. Bob uses a bagged potting mix, which is easiest to cart down onto the docks. But whether you buy by the bag or truck load, he says, "You want one that has lots of pumice in it for good drainage." He doesn't like mixes with moisture-retention beads. "All they do is float to the surface and grow algae."

Amend the mix. Minerals are often overlooked in planting soils. Bob adds amendments at the rate of one bag of potting soil (2 cubic feet) to 4 cups

of amendments. He likes 1 cup *each* of dolomite, azomite, greensand, and diatomaceous earth. He also feeds regularly, following fertilizer instructions, and delivers an inch of water to each pot when the surface soil dries out.

Renew your soil. If the soil has sunk down in the container and appears compacted, Bob revives it by dumping out a container onto a tarp, removing the plants, and mixing together half old and half new potting soil. To this he will amend in the same proportions as for new soil.

Try zonal experiments. You may be surprised. Although Bob lives in Seattle's relatively mild USDA Zone 8b, the lake can be subject to

Contained exuberance. Tall angel's trumpets (*Brugmansia spp.*) anchor a lively grouping of container plants that appears to be growing right out of the deck beside Bob Lilly's houseboat wall. Red blooms of flowering maple (*Abutilon* 'Nabob') and hybrid geranium brighten the center, while silvery *Artemisia* 'Powis Castle' and *Dichondra argentea* 'Silver Falls' lace along the deck's edge. Pots of silver echeveria succulents perform as finials on the dock posts.

Plants appear to grow on water. The lush foliage of a rollicking collection of container plants spills down to Lake Union's waters, completely shrouding the decking of Bob Lilly's houseboat.

what are known as "Canadian blasts," and he says, "I've had containers freeze solid—lilies to 20 below." Pines, junipers, wisteria, and sempervivums have also made it through. "But of course," he admits, "I've lost a lot."

Be bold with your plantings. Win or lose, Bob will try growing just about anything in a pot. He discovered orienpet hybrids *Lilium* 'Scheherazade' and *L.* 'Silk Road' were not good candidates because "the bulbs grow so fast, they get to the edge and bruise themselves." But that leaves a lot of other lilies to grow—more than sixty varieties appear on the docks. And they can stay in their containers for a surprising amount of

time. "I had eight to ten 'Uchida' (*L. speciosum* var. 'Uchida') in a large wooden pot for sixteen years," Bob says. "When I took it apart, there were at least fifty bulbs in it."

Plant what you love. With space at a premium, Bob goes for tall and dramatic.

- *Brugmansia* 'Charles Grimaldi': Bob's favorite angel's trumpet.

- Italian shell bean, Tongue of Fire (*Borlotto Lingua di Fuoco*): "I love the color on eight-foot-tall bamboo poles," Bob says.

- Kangaroo apple (*Solanum aviculare*): Bob says, "It has constant flowers, and its attractive green fruit turns shiny orange."

- Kiss-me-over-the-garden-gate (*Persicaria orientalis*): Bob grows this annual every year. It can top out at ten feet.

- White four o'clocks (*Mirabilis jalapa*): "Fabulous," Bob says. "They get so big—twenty-two inches tall, in large containers—they're over my head."

Bob Lilly's container gardens surround his houseboat as well as several others.

Grow anywhere you find space. Container gardens extend beyond Bob's own house. Here, at a wide spot on the dock, he has grouped a variety of plants. They include tall pokeweed (*Phytolacca americana*) whose large green leaves contrast with the intensely fragrant white four o'clock flowers (*Mirabilis jalapa*) at right. Below, light blue agapanthus blooms next to orange echinacea, red big-leaf begonia, white gaura (*Gaura lindheimeri*), and orange and pink calibrachoa. Behind, golden bamboo and coastal daisy tree (*Olearia solandri*) rise up over the dock.

Why Grow a Healthy Garden?

In this time of upheaval and rapid change, how we interact with our gardens matters more than ever. These methods for increasing the health of the flora and the fauna around you are your contributions to the natural world we live in, no matter how small or big they may be. All things growing are pieces in a puzzle that are intertwined in ways we are just beginning to understand. Whether you decide to care for a single houseplant or acres of garden, you are connecting yourself to a world of wisdom that is far greater than you can know, full of mystery, surprises, and delight. You become the tiniest part of the massive, fragile, and yet resilient web that lives and grows and passes on. Welcome home.

Harvest the rain. Where rainfall is scarce and droughts are recurring, as is the case west of America's continental divide, gardeners must sometimes rely only on rainfall to irrigate their plants. That's why the owner of this New Mexico garden built a trough on his roof to capture and channel any rainfall to where it is needed most—the garden below. Rainwater spills from the trough to the dry creek bed below, and from there, to the plants.

HEALTHY YOU

"I believe the land is our teacher, our nurturer and provider in every sense of human experience. It is up to us to build a relationship to benefit from what the land can offer—whether it be food or medicine, beauty, spiritual or psychological benefits, or just a place to feel safe and human."

—Jessi Bloom, garden author

Go stylish with stone. If you live in an area where humidity is high and where untreated wood can rot quickly, consider using stacked stone instead. The beds pictured here, built of rock and dry-stacked flagstone, replaced a water-guzzling front lawn. The largest stones were placed around the bed's the bottom, with narrower flagstones above, An added benefit for cold-climate areas, stone retains heat longer than wood, making it especially great for growing tomatoes. Crushed rock covers the paths between the beds, while an urn fountain burbles softly in the center.

The Health Benefits of Gardens

In Part One, we focused on designing, planting, and maintaining healthy gardens, no matter how big or small yours happens to be. In other words, methods to boost your garden's health. You work hard to make your garden healthy, so now let's ask—what does your garden do to boost your health?

Many gardeners have already discovered their gardens' gifts, and they're as varied as their gardens. Charlotte Ekker Wiggins, a quilter in Rolla, Missouri, says her garden enlivens her and her business. "My garden is where I recharge my spirit, where I get inspiration for my custom-quilt business, and where I am reminded that we are all connected."

Journalist Erica Browne Grivas recognizes a multitude of gifts from her Seattle, Washington, garden. She says, "My garden offers me a laboratory to work off the stresses of the day, and a canvas to practice creativity, along with excessive gifts of food, flowers, and seeds."

Sometimes aesthetics combine with sustenance. That's the case with Stephanie Niedermyer, who grows twenty-six different kinds of beans in her small backyard in Eugene, Oregon. She says, "I'm a magpie, attracted to pretty, bright things—that's beans!"

Vanca Lumsden, a garden artist and self-described plant addict, jokes about her gardening experience, "I owe my sanity and my children owe their lives to the fact that I garden."

And author and humorist Carol Michel, who gardens in suburban Indianapolis, Indiana, says, "My garden makes me work, sweat, and occasionally swear. But just when I think I should give it up for some tamer pursuit—like mountain climbing—it offers me a lovely scented bloom, or a ripe tomato, or a peaceful moment of calm, and I decide it's worth tending it after all."

In this section, we'll explore why "getting dirty" is good for you. We'll look at the gifts that gardens provide—gifts for the body and for the mind. Certainly, you will derive sustenance from the food you can grow. Whether it's a few peppers in a pot, a row of garlic tucked into the flowers, or a dedicated vegetable patch, what you harvest yourself is some of the healthiest foods you can consume. And the tastiest. These vegetables have traveled the shortest distance possible to end up on your plate. As one home gardener noted, "This carrot doesn't know it's been picked yet."

Gardens also provide exercise that benefits your overall health. Do you know how many calories you burn loading a wheelbarrow with weeds? Or how much you can build and maintain strength by mowing your lawn with a push mower, or digging the soil in a planting bed? Your ten thousand steps can add up fast when you're out caring for the vegetables and flowers.

Gardens are profligate when it comes to gifts for the mind. Outdoor spaces enhance creativity. Their beauty—whether it's from roses in full bloom, or fall leaves backlit by a setting sun—continues to inspire artists, musicians, and authors. Even the process of creating a garden provides a sense of accomplishment and a soothing link to nature. Gardens reduce stress and offer relaxation. Gardens give us places to play, to make a solitary retreat, or to connect with family and friends.

So, why does our relationship with the natural world matter? Quite simply, we humans are hard-wired for it. Science now supports the idea that nature's outsides are good for our insides. "We were meant to be outdoors, and at one with nature," says Dr. Tanya Hudson, an Oregon-based naturopathic specialist. "And although we may not realize it, when we're outdoors in nature, we activate all our senses," she says. "Scents, the feeling of breezes on our skin, all help us to practice mindfulness. Time spent outdoors, in nature, also reduces stress and boosts energy."

Build a Stumpery. One of the best ways to display ferns and other plants that thrive in shady, moist areas is to build a small stumpery—a garden feature similar to a rock garden but built from whole tree stumps or fallen logs. For the one pictured here, in a Pacific Northwest garden, the owners had a fallen tree's root system brought in and positioned on its side. Then they built a garden around it. Sword ferns and mosses cloak the stump, while a low, horizontal May apple (*Podophyllum* 'Spotty Dotty') grows in front. A Tasmanian tree fern (*Dicksonia antarctica*) marks a curve in the mossy path, behind.

Why Grow Your Own Food?
The Gift of Sustenance

Healthy crops are grown in healthy soil and matured without chemical interference. First, we heard from Hippocrates: "Let your food be your medicine." Then, from our grandmothers: "Eat your vegetables; they're good for you." And now, from seed grower Rene Shepherd: "The more color on your plate, the healthier it is."

All of them are right. As you eat more fruit and veggies, your risk of heart disease, cancer, and stroke plummets. And when you grow what you eat, you get incomparably sweet, vine-ripened produce. Fresh really is best.

All this good news isn't lost on plant breeders, who are developing vegetables that contain everything from more vitamins to extra antioxidants. And the availability of heirloom vegetables, whose seed has been saved by generations of farmers and gardeners, has never been greater. By growing fresh produce right outside your back door, you'll be more inclined to put it on your plate.

All this is not lost on health care professionals, either. Many berries, especially blueberries, are often called "superfoods" because they contain potent antioxidants and help to increase the body's level of good cholesterol (HDL). "Blueberries contain lots of vitamin C," says Dr. Tanya, so they're "excellent for cognition." They're also easy to eat. For a healthy snack, toss a handful of blueberries into smoothies or yogurt. Cabbage, carrots, kale, spinach, and tomatoes are other healthy choices. Herbs like parsley provide vitamins and minerals, and they add fresh flavor to foods without adding calories. Try mixing whole leaves of basil, cilantro, or parsley into salads.

"Healthy food is like medicine, adds Dr. Tanya. "You *are* what you eat. Fiber is important for your gut. And once your gut improves, so does your brain. It's a beautiful continuum—from your garden to a healthier digestive system, to a healthier brain." The best part, she says: "Your taste buds light up when they can taste the true flavors from the earth."

Previous page and right: **Build in sustainability.**
If you have room, consider making
your raised beds the focal point
of your back yard. Then build in
sustainability features around them.
For the small urban farm pictured
here, in Seattle, Washington, the
owner maximized growing space
by tucking smaller beds into tight
spaces and trellises around the
garden's perimeter to support
vining food crops and provide
privacy from neighbors. She added
chicken coops and birdhouses.
Well-placed cisterns collect and
store rainwater; each connects to
a garden hose for irrigation when
needed. Paths are covered with
permeable wood chip mulch.

Plan for Success

Starting a vegetable garden can be daunting. Even though there are far fewer choices for vegetables than there are for ornamentals, concerns about cool-season vs. warm-season vegetables, succession planting, and crop rotation makes the whole design and decision process feel intimidating. The best advice when starting out—figure out your options.

Light: Begin by selecting the best spot on your property for growing vegetables. Perhaps that's the little bed that edges your side yard. Or that patch of soil that frames your back lawn. Or

No space? No problem. Grow edibles in containers to arrange in stylish groupings on a patio. That's how food gardening expert and author Rosalind Creasy manages to grow crops in every nook and cranny of her Northern California property—including near the front steps. Tuck lettuces in three to five smaller pots, twelve to fourteen inches in diameter.

those boxes on your deck. You can also tuck edibles in among your existing ornamentals. Most vegetables need a minimum of six to eight hours of direct sunlight. A few, such as parsley, leaf lettuce, arugula, cilantro, beets, and radishes can get by with as little as four hours of direct light. That might make them candidates for containers on an overhung balcony where the sun moves through at a certain time of day.

Soil: Once you have located your sunniest spot, consider its soil; whether it's rich loam or heavy clay could help determine how best to plant and grow your crops. If your soil is not terrific, or your space is limited, you may choose to grow in raised beds or large well-draining containers, or even fabric grow bags. Or grow both in the ground and in containers, depending on the vegetables and the space you have.

Amendments: Whether you choose to grow your crops in the ground or above it, good compost is your best amendment. It keeps your soil loose and easy for roots to grow through. For raised beds and containers, you can choose a potting mix that will provide optimum growing conditions. A two-inch mulch around everything holds moisture and keeps down weeds.

Water: Most food crops need a steady supply. In parts of the country blessed with year-round rainfall, a hose can be enough to deliver supplemental water in dry times. You can also run a soaker hose along the vegetables, but these don't deliver water evenly—more at the front end, less as you go down the line.

Drip irrigation will solve that problem, since the emitters are engineered for the same delivery across the whole length of pipe. For drier parts of the country with limited rainfall, a simple system, coupled with a battery-powered hose-bib timer will keep your plants growing well. An important note—find a brand that's readily available and stick with it for all the bits and pieces. Like train tracks in the Old West that ran on different gauges, there is no standardization of drip piping. A half-inch plastic pipe of one brand will not properly fit the couplings from another brand.

The easiest way to set up a drip system is to buy a kit that comes with all you need to water a specific area, usually measured in square feet. Water is delivered through preset emitters punched six or twelve inches apart in the tubing. In most kits, the emitters each put out a gallon of water an hour. With those numbers, you can figure out how long to water. After you lay out the pipes, fine tune how much water to use with careful observation—digging down near the roots of a plant after you water to see where the moisture is going, and noticing any wilting leaves. And how much water you use changes throughout the season. In hotter weather, you'll raise the watering times to keep up with your plants' needs.

Move the harvest, easily. "Shopping" for dinner is especially easy for this garden's owners, BJ and Primus St. John. BJ simply picks the ripest crops, whether chard, tomatoes, squash, or salad greens. Then she passes them through the open "doors" for Primus to load in his basket. Those crops can determine their dinner menu. Here, fresh really is best.

Opposite: **Protect your beds.** If you live in deer country, protect your ripening crops with fences, as the owners of this garden in Washington State have done. They surrounded their custom-designed wood-framed raised beds with stylish fencing. The deer won't jump it, but the owners can access the raised beds through a gate at one end.

Brie Arthur

FOODSCAPING AND THE FIVE EASY VEGETABLES

Author and horticulturist Brie Arthur foresees a bountiful world of delicious possibilities in the traditional ornamental landscapes around homes. Among those permanent plantings of flowers and shrubs, Brie would like to see a mix of vegetables and grains, always cycling through the seasons while delivering delicious crops—without depending on markets far away.

"People have been looking at bare shelves in the stores," Brie says, "and realizing that one disturbance, like the COVID virus, could really impact food supplies." Not that your goal is to plant everything you eat. But it's yet another reason, Brie says, to boost your growing game with foodscaping. "Employ creative strategies for where to put your vegetables," she says, "and your neighbors might not even realize what you're doing. A lot of vegetables are amazingly ornamental."

Nurtured in gardening by her parents and grandparents, Brie earned a landscape design and horticulture degree from Purdue University, pursuing a career first in a public garden and then in several private nurseries. Now she's writing and speaking about her passion—foodscaping. Here's Brie's advice.

Ditch the rows. "You don't need them if you're not a farmer," Brie says. "Rows are for farm machinery. Planting and harvesting vegetables don't have to be messy." Vegetable gardens don't need to be segregated at the back of the house. If your ornamental beds and borders get six to eight hours of sun, that's the place to slip in vegetables. Brie likes to edge her beds and borders with a row of pepper plants. "Right up front where it's easy to harvest them," she says.

Grow vegetables in the vacancies. Brie says, "It's much easier to use space you've

Try a variety of containers. Brie experiments with any kind that drains well. Here, Brie puts large round fabric growbags and a yellow plastic bag that once held compost alongside traditional glazed pottery to complete the eclectic grouping in her above-ground vegetable garden.

already cultivated. And some neighborhood covenants prevent you from digging up any more turf." Stroll around your garden beds and identify empty spots covered in plain mulch between shrubs or perennials. "Put eggplants into the places where you might otherwise have planted summer annuals," she says. "Or in the cooler months, tuck lettuce or arugula in between permanent plants to reduce weed pressure."

Use grow bags. Brie is always on the lookout for products that make growing food easier. "There's this new fabric bag, made from recycled water bottles, that holds a hundred gallons of soil," she says. The bags come in all sizes—Brie's measures thirty-eight inches across and twenty inches tall. That gives her eight square feet of growing space. "It's like a raised bed, and inexpensive, compared to lumber. Last spring, I grew fifty heads of broccoli there—best harvest of broccoli I've had in my life."

Try Brie's five easy vegetables. "Start out with garlic instead of tomatoes, if you're a novice with growing vegetables," Brie says. Tomatoes can be challenging. "If tomatoes are the first thing you try, you might get your feelings hurt and quit." But if one of Brie's five easy edibles isn't to your taste, try something else, easy or not, that you know you'll like to eat. You're more likely to stick with a vegetable you really care about, even if growing it presents you with conundrums.

- **Arugula:** Easiest green of all. The leaves flourish in cool weather. It self-seeds, but Brie believes the ones that come up on their own are even tastier, as if they've settled right in to your place. Arugula makes a great groundcover between other plants.

- **Garlic:** "Garlic is the no-fail vegetable," Brie says. Plant garlic bulbs in fall after the weather and soil have cooled, and harvest the following July. Edge a border with garlic for an upright accent.

- **Onion:** Grow from seed or sets—small-size bulbs. Sets produce onions more quickly, which is good in cooler climates. Plant in early spring when soil can be worked. In warmer places, both seed and sets will give good results. The vertical foliage contrasts well among plants with mounding shapes.

- **Peppers:** The easiest, and arguably the most handsome, of the warm-season vegetables, both hot and sweet peppers make a beautiful addition to your sunny garden. Try growing them in the front of the border as an edger, like Brie does.

- **Potatoes:** There are many methods for planting potatoes, but in early spring, make sure your soil has warmed to forty-five degrees. Harvest new potatoes after flowering, and the main crop after the foliage dies down. Handsome leaves and tiny flowers make potatoes ornamental throughout the season.

Find out more about Brie Arthur and her book The Foodscape Revolution *at her website, www.briegrows.com, and look for her garden advice on her YouTube channel, Brie, the Plant Lady.*

Shop smart. Before heading to your nursery, measure the length and width of the bed you've set aside for your crops. Write down those measurements so you can figure out roughly the number of crops you can fit in that space. Also make a list of the crops you would like to grow. At the nursery, check plant labels; note especially the approximate size each crop will be at maturity before loading them on your wagon.

Seeds or Seedlings?

Start with the method that's easiest for you. Often, buying six-packs at your favorite independent garden center or by mail order is simpler. While your choice of varieties is fewer than if you buy seeds, you'll be able to space your plants properly, and you get a head start on the growing season.

On the other hand, some plants, such as arugula, green beans, peas, swiss chard, and spinach, are very easy to grow from seed. Another plus: You can extend the harvest season by a month or more by continuing to sow seeds every two weeks. Or you can grow your seedlings in six packs, small pots, or rectangular flats. The advantage of making your own starts is that you get the best of both worlds—the varieties you want, and the ability to more closely monitor germination.

Give your plants homecoming care.

Place your nursery purchases in a shady area; water them well. If possible, set aside an area as your permanent spot for planting tasks, like the one pictured here. Trellised grapevines shade its vintage table, used for potting. A thick layer of mulch provides a permeable surface on the ground around it.

How to Plant Seeds

Follow the instructions on the seed packet about when to sow, how far apart, and how deep to plant. Some small seeds (beets, lettuce) are also sold as seed tapes with the seeds glued to a biodegradable tape for easy spacing. Cool-season plants can usually be sown as soon as the ground can be worked in spring. Warm-season seeds need soil temperatures around 55 degrees.

Plan out your planting. Before planting, set your new seedlings atop the soil surface; adjust spacing between them, as needed, to allow for the plants' future growth. Erect trellises to support any vining crops, such as beans. These tripod-style trellises are made of narrow bamboo stakes tied together with jute. Settle plants into the soil, firm the soil around root balls, then water well.

Start your seeds in clean six-pack cells, two-inch pots, shallow flats, or even egg cartons. It's all right to use recycled containers, as long as you wash them out. For an extra precaution, you can soak them in a 10 percent bleach solution and rinse well. Fill them with a seed-starting mix and cover the seeds with the mix or sand to the appropriate depth. The biggest cause of failure to sprout? Seeds are planted too deep. Go easy. If in doubt, sprinkle on a thinner covering.

Indoors, you can get your seeds started in a tray on a sunny warm windowsill, or on a sheltered back porch. There's also a whole range of indoor lighting kits and planting tables if this appeals to you. Start modestly to see if it fits your lifestyle when it comes to tending the plants. This is a whole world of equipment, with specialized water-holding containers, timers, heat mats, and different lights, from fluorescent to the newer LEDS. The world opens up even wider with covered frost-proof boxes and greenhouses. These items may be just your ticket for getting your seeds off to a great start, but when you're beginning, find out what your exact needs are before diving in. The windowsill or a single grow light may suit you perfectly.

Recycle egg cartons. Paper-pulp egg cartons make ideal trays for starting small seeds. Simply fill each cup with a light soil potting mix, then tuck one or two seeds into each cup. Put your seed tray in a warm light spot, such as near a bright window. Mist the soil regularly. Transplant the starts into garden beds after the first few leaves appear.

Gardening by the Numbers

So, how many vegetables should you plant? Ed Hume, award-winning TV and radio garden personality and founder of his family-owned business, Hume Seeds, suggests planning first on paper. He says, "Your plan might include decisions on family needs, lists of your favorite varieties, and the number of plants or space devoted to each crop."

His suggestions for how many plants you need are based on a family of four. "You can adjust the numbers," Ed says, "depending on how many people you'll be feeding, their appetites, and preferences." Seed packets are generous. Share with friends and neighbors if you have more than you can use.

Cool-season:

- Broccoli (12–15 plants): Sow indoors at ⅛ inch depth; move outside approximately a month after germination.

- Cabbage (12–15 plants): Sow at ⅛ inch. Light-weight row covers are effective to block cabbage moths.

- Cauliflower (12–15 plants): Sow at ⅛ inch. This one likes to be "moderately moist," Ed says.

- Lettuce (leaf) (30 plants): Sow at ⅛ inch. Make sowings every two weeks until midspring, then again in fall.

- Kohlrabi (15 plants): Sow at ⅛ inch. "Excellent with dips," Ed says.

- Spinach (10–20 plants): Sow at ¼ inch.

Warm-season:

- Peppers (6–10 plants): Sow at ⅛ to ¼ inch, depending on variety.

- Pumpkins (3 plants): Sow at ½ inch depth. Ed says this is the "ideal Halloween crop."

- Squash (3–6 plants): Sow at ½ inch depth. So many kinds, both summer and winter.

- Tomatoes (10–15 plants) Sow at ⅛ inch depth. Choice of cherry, slicing, or sauce tomatoes.

Pat Munts

SHORT-SEASON STRATEGIES FOR SUCCESSFUL HARVESTS

"Ninety days is what I've got to work with," says Pat Munts, extension and urban horticulture coordinator for Washington State University. Pat lives in Spokane, Washington, east of the Cascade Mountains and in the foothills of the Rockies. "We can get occasional frost in July and August," Pat says. "Forty-five-degree nights are quite common."

With that kind of truncated timeline, Pat has developed strategies that allow gardeners to maximize their growing season. If you live in the northern tier of the US with similar conditions, or if you're farther south and just want to extend your garden's productivity through periods of frost, try some of Pat's suggestions.

Choose the right varieties. Read the fine print on seed packets and plant labels. Maturity rates for the same vegetable may differ, depending on which variety you choose.

Peppers, for example, can ripen any time between 55 to 120 days from the date you plant them. "Read the tag first," Pat advises, "and then plant accordingly."

Avoid rushing your plants. Don't force warm-season veggies into cold ground, Pat says. "A short-season gardener's best tool is a soil thermometer. "Crops such as eggplant and squash need soil that stays close to sixty degrees. Warm up your soil by covering it with black plastic, or better yet, with a green plastic sheeting sold as solar mulch," Pat suggests.. The solar mulch absorbs infrared light. "I took my soil from sixty to seventy degrees in ten days," Pat says. You could warm the soil and then remove the sheeting for use in subsequent years. To retain the heat you've gained, Pat says, "Make sure you put down a good, thick—at least two-inch—layer of mulch."

Meet plants' needs. As plants race to grow and produce food, it's important to provide soil nutrients. "No plant can produce well if it's stunted," Pat says. "Make sure you fertilize as needed and water plants regularly." If you must feed, try an organic fertilizer with numbers like 3-5-4. For a healthy boost in containers, some plant foods have additional mycorrhizae, those fungi that establish symbiotic relationships with plant roots.

Organize frost protection now. Hustling around for floating row covers or old quilts when temperatures are predicted to drop later that night is no fun. Gather your materials before snows or frost descend. "In short seasons," Pat says, "night temperatures start to drop even when the days that precede them are warm. Temporary coverage can get vegetables to ripen." Where the growing time is even shorter and nights routinely go below 55 degrees, Pat says, "Folks simply grow their warm-season vegetables in greenhouses or tall poly tunnels." Row covers range from the lightest weight—good for keeping out insects—to the heaviest winter weight, which also blocks the most sunlight. Frost protection vs. light needs? "It's a tradeoff," Pat says. She prefers a middle-weight cover.

Harvest with care. If a hard frost is predicted, pick and store produce. Cantaloupe, apples, and tomatoes will continue to ripen. Eggplants, peppers, and squash, on the other hand, will not ripen any more—the stage when you pick them is what you get. Handle your crops gently. "A little bit of damage to the skin can rot them quickly," Pat says. Another way to go—pull out a whole tomato vine, hang it in a frost-free place, and the fruit will color up. "The leaves make a mess in your garage," says Pat, "but it works."

Pat Munts has been gardening outside Spokane for forty-two years. She is the coauthor of The Northwest Gardener's Handbook: Your Complete Guide.

Protection means a longer growing time. Row covers and a greenhouse are essentials for short-season gardeners Jennifer and Lance Barker, who live at an elevation of 5,000 feet in eastern Oregon. The couple grow most of their food, much of it under cover, which allows year-round harvests.

Foil pests. Small raised beds that sit well above the ground are ideal for growing lettuce, which all kinds of snails, slugs, bugs, and animals find delectable. This bed is designed to stop those critters and to provide extra protection against frost on cold nights. A sheet of plexiglass, attached to its back with hinges, acts as a cover that's easy to pull down and lay flat, well above the tops of ripening crops.

Opposite: **Add blooms.** Tuck pollinator plants between crops, for additional benefits. As bees flit among blooms sipping nectar from crops such as the coreopsis, flowering thyme, and tiny daisies, they'll pause to explore on your ripening crops.

When and How to Harvest

Figuring out when and how to pick your produce can be another learning process. Go in with an experimental mindset, especially when growing a new (to you) vegetable. Do those beans taste better when picked in their skinny stage, or does the flavor develop better when they're more robust? Pluck a bean off the vine at various growth stages to determine what's best—for you, and for that kind of bean.

If possible, harvest herbs in the early morning, when their leaves and stems carry the most water. But right before an evening meal is the perfect time to look around your garden for what crops to include on your dinner plate. If you can, don't hurry as if you're grabbing produce from a grocery store bin. Instead, consider approaching your harvesting with gratitude. That response, says Jessi Bloom, author of the book *Creating Sanctuary*, "is common practice in herbalism and indigenous cultures." By acknowledging what the forces in nature have done to present you with this food, you find your humble connection in the greater growing world around you.

The quickest plants from seed to harvest are radishes, and if you've never tasted homegrown radishes, you're in for a treat. With a variety of flavors from radishes like Starburst, Easter Egg II, and Crunchy Crimson, they are a fast, easy, and satisfying crop.

Plant some spillers. If possible, tuck a few nasturtiums around the edges of your raised beds. Trailing kinds can take over a small bed, but compact kinds that top out at eighteen inches are ideal for this use. You can use their blossoms to add color to salads, and their seeds have a peppery flavor.

Cut-and-come-again plants, like arugula, kale, leaf lettuce, mizuna, spinach, and other leafy greens make it easy to keep growing and harvesting from the same plants. "Harvesting only the outer leaves gives us a longer lettuce season," says Mark Turner, a garden photographer who grows greens under winter cover in Bellingham, Washington, "and with a few leaves from several different varieties, we have the opportunity to eat a more colorful salad."

If you're growing regular-sized eggplants, like the kind commonly found in the market, the fruit can become so heavy that it pulls the plant over. Surrounding the plants with low cages (suitable for peppers too) can help keep the fruit off the ground. But there are newer eggplant varieties, like Patio Baby and Millionaire, with small rounded or long thin eggplants whose skin never gets bitter, and the flowers and fruit keep on coming throughout the season.

All peppers, even sweet green ones, turn red (or orange or purple) if left on the plant long enough. So don't worry if you planted what you thought would be a green pepper, and now it's slowly turning red. It will taste the same—maybe even better.

"Harvest garlic when half of the leaves have turned yellow or brown and half are still green," suggests Peggy Riccio, owner of pegplant.com, an online resource for gardening in the Washington, DC, metro area. She also has a tip about watermelons. "Heavy rains may dilute the sugar content in the fruit, so leave a ripe melon on the vine for a few dry days to re-concentrate the sugar."

Is it possible to bring in a bumper crop of tomatoes before disease, critters, or skin cracking messes up your harvest? It's all about timing. Robert Pavlis, author of the GardenMyths.com blog, has made an intensive study of how to gather your tomatoes. When green tomatoes reach their full proportions, the stem starts sealing itself off from the fruit. After that, the tomatoes arrive at what's called the "breaker stage," when they have as little as 10 to 30 percent color.

"Once the breaker stage is reached," says Robert, "the fruit no longer relies on receiving nutrients or sugars from the plant, because it now contains all the ingredients it needs to fully ripen on its own." That's the moment to pick them, put them on your kitchen counter, and watch them finish turning red, or gold, or purple. Turns out, there's no perceptible difference in taste between a tomato picked at the breaker stage and ripened indoors and one left to "vine ripen." No difference except that you get to harvest more intact tomatoes. Thanks, Robert.

Harvest tomatoes. Most tomato plants are vigorous. Keep a wide, sturdy basket handy for picking and transporting your tomatoes to the kitchen. Check your plants daily, and pick any small fruits that are fully colored and large ones in the breaker stage. Doing so will extend your harvest season. But near season's end, when frosts are predicted, harvest all the fruit and store it in a slightly warmer place out of direct sun; check regularly for ripeness.

Harvest trick. To simplify your harvesting tasks, keep a colander handy so you can grab it, head to the garden, pick your salad greens, then wash and toss them before heading indoors.

Valerie Rice

GROW CROPS FOR FRESHNESS AND FLAVOR—SECRETS FROM A COOKBOOK AUTHOR AND MASTER GARDENER

Valerie Rice was just five years old when she discovered the magic of a garden and the crops that could come from it. While on a trip to visit family in Belgium, she helped her *Maake* (grandmother) harvest vegetables and then cook a delicious dinner using those crops. "I was raised in Newport Beach, California," Val says, "and until that moment, I thought that all food came from a grocery store shelf." That's when the seed of her life's work was planted.

Val honed her skills gradually through her school and college years. She settled in Southern California and now has a kitchen garden that would make her *Maake* proud, where she grows her own season-to-season crops. In her walled garden, three four-by-eight-foot raised beds, along with containers and in-ground planting pockets, burst with crops in every season.

The best part: She can grow vegetables that are often hard to find in markets. Among her favorites: lemon cucumbers, Mexican zucchinis, padron peppers, romanesco (also known as Roman cauliflower), and watermelon radish. And she can pick dinner and serve it fresh to her family and friends. Why is fresh, best? The longer that crops have been off the plant, and the farther their journey to reach your kitchen, Val says, the more depleted of nutrients and flavor they are likely to be.

Her mild coastal climate allows planting cool season crops in fall (alpine strawberries, fava beans, lettuce, new potatoes, peas, radishes, and rhubarb) for harvest winter into spring, when she refreshes her soil and plants warm season crops such as tomatoes for summer harvest. But whether your growing season is short or long, think of your garden as your backyard farmers' market. "The garden is my starting point for every meal," Val says. "It takes the guesswork out of menu planning. It has pushed me to be a more creative cook and gardener."

Here's Val's advice.

Give plant roots the room they need. Some plants, including herbs, can grow and thrive in beds that are six inches deep. But as Val has discovered, deeper is better for most other crops. Val's four-by-eight-foot raised beds now measure eighteen inches deep.

Tend the soil regularly. Amend the soil between seasonal plantings. Val digs in some homemade compost and an organic amendment containing worm castings and mycorrhizae.

Serve crops fresh. Many of Val's "Pick-Mix-Serve" recipes have been inspired by restaurant meals, featuring just-picked crops. "They bring travel to your table," she says. For example, here's a fresh salad, from Florence, Italy, that you can put together quickly: "Shave a clean, raw zucchini into spaghetti-thin pieces; arrange them on a plate. Sprinkle with a good olive oil, and parmesan cheese."

Choose tomatoes for flavor. Among Val's favorite varieties are the following: 'Sun Gold' ("sweet orange fruits have superior flavor"); 'Brandywine' ("a delicious, big-fruited variety; great for salads"); 'Black Crème' ("the small fruits taste very floral"); 'San Marzano' ("it's indeterminate; I can harvest crops all summer"); 'Cherokee Purple'; and 'Green Zebra'.

Get creative with other crops. Wherever possible, tuck in a few crops that you've never before tried to grow, or ones that are new to you. Browse through nursery shelves or catalogs beforehand, then add those few surprises to your order. Then, at harvest time, experiment with them in your kitchen, as Val does. Some of her freshest recipe ideas came from "playing" in her kitchen. Among them:

- Arugula: "I use it in cocktails, in side dishes, even in juice in the morning; it's a cut-and-come-again crop."

- Blood orange: "I use it to make margaritas."

- Celery: "Braising it totally transforms it to a beautiful, delicious dish."

Valerie Rice is the author of Lush Life: Food and Drinks from the Garden. *Find out more at her blog, www.eat-drink-garden.com.*

Drinks are on the garden. Val harvests blossoms from pineapple guava (*Feijoa sellowiana*) to use in one of her many garden-fresh recipes. Their delicious flower petals can be used in salads or jelly and to make refreshing drinks. Pineapple guava grows best in warmer climates—from Hawaii and southern California to south Texas and Florida.

Raised beds give a beautiful lift. Raised beds four feet wide by eight feet long line up along paths for easy harvesting. Fruit trees, including olive and pineapple guava, grow among them. Containers filled with herbs are clustered just beyond. Val's work table is across the path from the beds, with a handy sink nearby and shelves for storing pots; it's perfect for seed starting and planting small containers.

Pick and serve a berry dessert. Like many sun-ripened fruits, blueberries and strawberries are at their juiciest and most flavorful when they've fully ripened on the plant. Pick those berries to rinse and serve as fresh desserts, and you'll be glad you did. If you live in warm a climate, choose heat-tolerant rabbiteye blueberry.

Opposite: **Serve your tomatoes fresh.** Slice them on salads and sandwiches and enjoy their full, ripened flavors. Which varieties are most flavorful? Experiment by growing several kinds over several seasons. 'Green Zebra' is on the tart side; small red 'Early Girl' is versatile and tasty; 'Black Krim' and 'Amana Orange' are heirlooms (great for saving seeds); and 'Brandywine' is delicious. Red-fruited varieties generally pack in the most nutrients, including the vitamins C and K, potassium, and lycopene (an antioxidant).

Save Seeds—or Not

The generosity with which plants develop their seeds makes it tempting to go ahead and save a few for future use. You can save those that show the traits you value—say, you want an earlier harvest, so you save seeds from the first corn that was ready, hoping those jack-rabbit traits will carry forward. Or you save seed from that arugula that survived the winter's cold when other greens succumbed. By doing this, you're saving seeds from plants that have grown in your location, and that have successfully adapted to your micro-climate and soil. Over time, each generation becomes more acclimatized to your place.

Seed saving makes sense if the plants you started with are "open pollinated." That's where the same stable traits of the parents are passed on to the offspring. Most heirloom plants, those that have been passed down from generation to generation, are open pollinated and stable. Seed packets that carry these traits are usually labeled as such, and many companies offer them.

The other kind of seeds come from hybrid plants. Hybrids are those with parents from two different types, crossed together. This kind of breeding creates sturdy plants with what is known as *hybrid vigor*. Crosses are often made to strengthen the traits wanted, such as resistance to disease. The seeds from the following generation of hybrid plants are not stable. They fall out into the traits of each parent. Some will be like one parent, some like another. And some will be mixed, but not exactly like the strong hybrid you originally planted. Seed saving from a hybrid is a crap shoot. You don't know what you'll get. You might enjoy the gamble. Others prefer to buy another pack of hybridized seeds, so they'll grow the vegetable with the traits advertised.

Plant an herb bed. You can grow herbs just about anywhere, in terra cotta pots, wooden boxes with holes drilled in their bottoms, or raised beds and flower borders. But if you're a serious cook and you have room in your garden, consider dedicating a bed to herbs. Low wattle fences, built of woven twigs, divide this herb bed into sections, each filled with a different set of herbs—from chives (upper left) and mints (left) to thymes (scattered throughout).

Make an herb bouquet. Create a bouquet of fresh herbs to display on your kitchen counter for a day or two. Then, pluck them fresh to use as garnish or to toss in salads. The sprigs in this bouquet include red- and green-leafed basil, mint, sage, and thyme (with flowers).

The Best Is Yet to Come

At season's end you may experience mixed harvest results. Yes, the peppers were tasty and bountiful, but some of the tomatoes were blighted and the cauliflower foliage looked more like lacework, caused by worms punching through it. Rabbits, deer, birds, raccoons, and squirrels also loved harvesting in their own way—spitting out your huge strawberries after a single bite. The weather was either a deluge or a drought. All those high hopes at the season's start might seem absurd in the rearview mirror.

But now is not the time to give up. No matter how discouraged you may be, don't take gardening disasters personally. Know that a bad season is not your fault—and don't get down on yourself for what you didn't know or couldn't do. There are many reasons crops fail, and much is beyond your control. The best gardeners know that every year is different. Those who garden year after year learn from mistakes, remember triumphs, and go forward. Be like those gardeners.

Maybe your ambitions were too high—next time, you'll pare it down. Maybe you had to learn about watering the hard way—seedlings might need sprinkling every day, or even twice a day in hot weather. Maybe you chose the wrong plants. Or you need to protect your young crops from hungry critters by covering them differently. Whatever setbacks occur, just keep on learning. Gardeners' knowledge is hard-won, but once you have it, you can build on it. And that, in itself, is a generous gift.

Exercise Benefits of Gardens

All gardens, even small ones, need regular tending to stay healthy and productive. That includes pulling up weeds, mowing grass, raking leaves into piles for your compost bin, as well as pruning, refreshing the soil with compost, or digging planting holes. All are strenuous physical tasks that you might prefer to hire a landscape professional to do for you. But consider this: such garden tasks can be good for you.

According to research gathered by experts over decades, all that bending, lifting, reaching, mowing, squatting, and pushing wheelbarrows loaded with soil can provide aerobic, isotonic, and isometric exercise. Together, these activities benefit your muscles and bones, while improving your strength, flexibility, balance, and endurance.

Choose tools that help you work out wisely. Mow the lawn (if it's small) using a push mower. Rake up leaves instead of blowing them toward your property's back fence with a leaf blower. Use a long-handled, taper-tipped warren hoe to trace rows in prepared soil before seeding. Pull weeds by hand, dig them up, or use a U-hoe (also called a stirrup or scuffle hoe) to dislodge any that grow through gravel. Or use a long-handled weeder that has a set of metal claws on one end of the handle; you can punch those claws into the soil around tap-rooted weeds, and extricate the whole plant while you stand up. By tackling such garden chores by hand, you will benefit the environment as well; no noise pollution, no blowing of dust or pollen or leaves.

Performed regularly, garden tasks can provide steady aerobic exercise. Digging holes and shoveling soil are big calorie burners (250 to 350 calories per hour). Another plus: digging and raking involve multiple muscle use. To protect your back from injury while lifting a shovel filled with heavy dirt, bend at the knee and step forward as you raise that shovel to dump the dirt elsewhere.

But before you tackle any heavy-duty gardening task, warm up your muscles. Digging a planting bed before warming up is like jumping into an exercise routine at the gym before warming up. Result? A pulled muscle or two, which can send you to a chair with an ice pack. Warm up slowly, as you would with any exercise program.

Start with light exercise—walking around the garden, sweeping the patio, or tidying containers— that allows muscles to warm up for ten to fifteen minutes. By doing this, your blood flow increases, and your available oxygen levels go up, putting less pressure on your heart. Your body gets more efficient at regulating your energy production. Your muscles become more

Burn some calories. Why go to a gym when you can work out as you create a garden that feeds your soul? The best part? Many chores can burn serious calories. According to the National Gardening Association, thirty minutes of planting seedlings burns 162 calories; weeding or planting trees burns 182 calories; and digging, spading, and tilling burns 202 calories.

elastic and your range of motion improves—all by spending a few minutes messing around with your plants.

Then when you do take on a more challenging job, change your position frequently, say, every twenty to thirty minutes. If your task will involve kneeling to plant a flower bed, tuck a plastic-covered kneeling pad made from thick, extra-large foam beneath your knees; Corona Tools makes a mighty one with memory foam. Or use a lightweight, portable kneeler bench to keep you well positioned. Stand up and stretch every ten to fifteen minutes.

It's easy to lose track of time when you're fully engaged in gardening. And before you know it, you may discover that you've been bending over that planting bed for six hours. Vary your gardening activities. Switch up tasks. (See Toni Gattone's advice on page 148.) Plan on taking breaks frequently throughout the day to drink water and rest, and if necessary, to reapply sunscreen (at least SPF 30) between jobs.

One last note about garden chores. Before you switch your attention to pruning roses, pull on a pair of elbow-length leather gloves. Even a single prick or scratch from a rose thorn can cause a potentially serious fungal infection called *sporotrichosis*. (The fungus that causes this is also associated with sphagnum moss, so protecting your hands for many garden activities is important.)

Relax. Set aside a place on your patio or beneath a shade tree where you can sit, rehydrate, and relax at the end of your gardening day, where you can survey those newly planted shrubs or perennials in your garden beds as you relax your tired muscles. For this patio garden in Malibu, California, a single chair positioned to face flower beds is especially inviting.

Toni Gattone

KEEP ON KEEPING ON—HOW TO BE A FOREVER GARDENER

Toni Gattone doesn't ever want to stop gardening. "We all have physical limitations," says this dynamic garden speaker and author of a book on adaptive gardening. "Especially as we get older. But the question is—how and when and what do you so you can keep on doing what you love?" The answer, Toni has found, comes by modifying gardening techniques and tools, as well as the actual work space.

"I live adaptive gardening," she says, due to the modifications she makes to accommodate her chronic back trouble. Whether you're new to gardening or have habits engrained from years of tending your own plot of land, Toni urges you to adapt yourself to your garden and adapt the garden to you. Here's her advice.

Change your work patterns. Instead of drilling down on a single project, work in smaller time periods. Perhaps you could follow the example of butterflies or bees. They flit from one flower to the next—never staying too long in one place. Tackle a series of smaller tasks—weed for fifteen minutes, then switch to seeding a couple of containers with lettuce, for example, and then go on to rake some soil into the lawn's low spots. That way you use different muscles for each job and run less chance of overworking.

Take breaks. Stop after completing one short task, such as clipping blooms from a basil plant. Don't rush right away to the next job on your list. Take a few minutes to sit, look around, enjoy the sensory experience of what's around you. What do you hear? What can you touch? Taste? That small conscious moment of stillness between projects can

Make adjustments to fit your exact needs. Toni always modifies for comfort. A regular two-foot-tall raised bed meant too much low bending, so, she raised it to the level that suits her perfectly.

bring your mind to rest and enliven your work. Then, plunge into the next task. If you break up an hour and a half of gardening into fifteen-minute task intervals, that will give you at least five opportunities to refresh yourself.

Adjust your work surfaces to your needs. "Most raised beds come up to my knees, which forces you to bend over or kneel down." Toni says. That's too low. She upgraded those beds with elevated boxes that reach just below her waist so she could easily work from a chair next to them or comfortably stand without a lot of bending. For accessibility, arrange your raised or elevated beds so they can be reached from both sides, each no more than four feet across—three feet across may be even better.

Alter your tools. To make some of her favorite tools easier to use, Toni added new features to them. For example, she slid a bicycle handlebar grip onto her skinny-handled trowel. "That made it larger and cushier and easier to grasp," she says. For a more comfortable grip on other tools, try dipping handles in a brightly colored, rubberized coating product called Easy Dip, which is available in hardware and other home stores. Rubber coatings make tools easier to hold, and their brilliant colors mean they'll never get lost in your garden. The product dries out after opening, so decant the dip into an airtight container if you want to use it again. Or, Toni suggests, "Dip all your handles at once."

Wear comfortable gloves. Different jobs require different kinds of gloves. Toni uses thin nitrile/nylon gloves for finer work and heavier gloves with goat or cowhide palms for repetitive tasks such as shoveling or raking. Lined waterproof gloves make cold winter tasks easier. And gauntlets that cover arms to the elbows are indispensable, Toni says, "for pruning roses, lemons, and berries. The older you get, those nicks on your arms take the longest time to heal."

By following Toni's tips, you can experience all your garden has to give. And, yes, she says, "You *can* garden for life."

Toni Gattone's book on adaptive gardening is The Lifelong Gardener: Garden with Ease and Joy at Any Age. *Find more about her road to resilience at her website, www.ToniGattone.com.*

Opposite: **Create a serene space.** If possible, locate it toward the back of your garden, and plant hedges or billowy shrubs around it. Add a bench there—if possible, facing a view or a birdbath. An opening in this copper beech hedge provides access to a tranquil space, set against a backdrop of shrubs. It's a perfect setting in which to decompress after a day's work. (No phones allowed.)

Gardening for Tranquility

Gardens are not just enclosures where you can grow plants. They are, or can be, sanctuaries from the outside world.

What gives a garden its tranquility? Trees or vine-covered fences for privacy? The sound of water, trickling in a small fountain, or the scent of jasmine wafting in the air? A weathered wooden bench, barely visible through an opening in a hedge? A path that meanders and then disappears beneath a rose-covered trellis? The tinkle of temple bells or wind chimes in a soft breeze?

Whatever your vision, a tranquil garden is a place where you'll want to sit quietly, if only for a moment, to connect with the natural world around you and savor your space. So it makes sense for you to make your garden uniquely yours.

Decorate your garden as you would a room indoors, by tossing colorful patterned pillows on a weathered bench. Or fill it with sweetly scented flowers. Find places to display objects that you love.

A tranquil garden is like a spa for your soul—a place where you can turn down the volume of your workday and tune up your senses while nature works its magic around you.

Savor springtime.
To build a garden is to create your own slice of nature, where you can relax and savor the fresh sights and scents that each season brings. Take time to pull out a lawn chair and put it where you can take in your garden's fresh sights and scents, especially in spring— whether a patch of tiny blue grape hyacinths as pictured here, the creamy white dogwood blossoms nearby, or the trees flecked with pink blossoms.

Provide shelter. Build a space that feeds your creativity and your soul. This shed serves as a creative space. (The owner is a writer.) Rain or shine, he can open the door to his garden.

Opposite: **Set the stage.** If space allows, set your shelter at the far end of your garden, and create a path that entices you and your guests to slow the journey there. For this garden, on Hawaii's Big Island, hand-carved Balinese sandstone statues of two lovers gazing at each other mark the start of the path leading to the pavilion. The owners and their guests place leis on these statues before heading to the pavilion, where they can sit and enjoy the views.

Add a Retreat

Open-air shelters, such as Southwest-style ramadas and Polynesian-style pavilions, provide a measure of comfort where you can sit or recline and find calm at the end of a day. Classic gazebos, usually built of wood and painted white, have a romantic quality.

You can buy ready-made garden structures, typically made from redwood or pressure-treated lumber. Explore your property to find the best spot, then stop and look around. Which vantage point provides the best views while blocking out anything unattractive? Whatever you choose for your garden's free-standing shelter, make sure it matches or blends with the style of your home and any other structures around it, such as fences or arbors, and complements your garden's style.

Enjoy the view. In this garden, the path's pavers create a diamond pattern that points toward the shapely garden bench at the far end, while the small fountain in the center is a focal point from both ends.

Opposite: **Hang a hammock.** Then climb into it, perhaps on a warm Saturday afternoon once chores are complete, and let the soft breezes and the sound of rustling leaves lull you to sleep.

Put Down a Path

If your space is big enough, be sure to include a path that connects the various outdoor living areas, especially if it wiggles through a meadow, or follows the trajectory of a dry stream bed, or disappears beneath a trellis.

Paths with permeable surfaces, such as gravel or bark, allow water to pass through them and are usually preferred in dry parts of the country. Impermeable paths made of solid paving or stone, on the other hand, are durable, easily accessible, and can include widely spaced pavers. Whatever material you choose, make your path wide for welcome or narrow for utility and casual strolls. Connect your walkway to a destination or circle it through your garden, leaving no dead ends. For example, a winding path can connect three patios—one for gathering, one for dining, and one containing a firepit area, so you'll enjoy your garden from three different vantage points.

Plant groundcovers or low tufting grasses or ferns near your path's edges. Add a focal point near a curve in your walkway, such as a small birdbath or a shapely stone that is lit from below as night falls. Then stroll along on soft summer evenings to see what's blooming or to savor the scents.

Tuck in an artful chair. Select a chair that fits your garden's style and the plants around it. Make it a destination—at the end of a path. That's what artist Linda Ernst did in the garden she calls "Dancing Ladies Garden," in Portland, Oregon. Ernst designed and made those three glass panels, mounted on the wall above the chair, to complement the blooms, including *Caryopteris* 'Sunshine Blue' in the foreground.

Perla Sofía Curbelo-Santiago

Invitation to touch. Helpful signs at the Buehler Enabling Garden invite visitors to reach out and feel the texture of different plants; signs also encourage you to plant a touch garden at home.

BRING HORTICULTURAL THERAPY HOME

"Ever since I was a little girl, I've loved to go pick flowers," says Perla Sofía Curbelo-Santiago, who comes from Arecibo, in the northern part of Puerto Rico. "I'm a real island girl." Growing up, she says, two major influences in her life combined—wellness and agriculture.

While in college, Perla explored the idea of majoring in environmental psychology, but ended up with a master's in communication. "Then," she says, "I found horticulture therapy, and I thought, oh, I can merge all my interests here."

Horticultural therapy, or HT, uses gardening and garden settings in conjunction with trained therapists to achieve treatment goals for both mind and body. HT helps

Easy access for all. Elevated beds at the Buehler Enabling Garden, part of the Chicago Botanic Garden, allow people to experience a broad range of plants and to interact with them using all the senses. The designs for pathways and borders they use here can easily be adapted for home gardens.

participants gain new skills or reclaim those that have been lost, such as strength or memory.

Perla completed her ten-month HT training at the Chicago Botanic Garden, which offers interactive and immersive certificate programs in a variety of subjects. She returned to Puerto Rico to spread the word about the health benefits of interacting with the natural world. Her work is now part of a proliferation of garden wellness programs in Central and South America, and she is contributing to the rise of this information in Spanish.

Perla is still going through the neighborhood, looking for flowers—but now she says, "I'm a wellness promoter, who uses gardening as a tool." Here are her ideas for bringing more peace of mind into your garden.

Consider your gardening goals. "HT gave me the framework of intention. That has been key for me," Perla says. Finding intentions is a thoughtful process. What state of mind do you want your garden to help you achieve? It could be comfort, it could be safety, it could be abundance. It could be as simple as one home owner who plants only one edible—potatoes—because it makes her feel rich when she digs up her harvest. Perla uses intention in her own garden. "I set goals such as a feeling of calmness. Then I do what I can to create a space that helps me get there."

Develop four seasons of sensory stimulation. Healing gardens involve all the senses, including smell, touch, and taste, as well as sight. Look for plants for your own space with color, texture, and fragrance. Place them within easy reach, either by a pathway, a tall container, or in raised beds. And don't forget that memory is also a sense. Which plants evoke your own memories?

Make your space accessible. HT gardens are always accessible for a wide range of physical abilities. Even if you and your family can easily traverse your garden now, designing for other wheeled modes of transportation, from wheelchairs to tricycles, makes sense. Smooth pathways and ramps, for instance, may be inviting for visitors and family alike, and will also assist you when you're hauling a loaded wheelbarrow around.

Think about comfort. Therapy gardens encourage participants to freely interact with the natural world. But sometimes that world can be harsh. Provide relief from hammering sun with benches or chairs under trees or other overhead structures. For the colder times, create windbreaks from trees, fences, or hedges and tuck seating areas near them. The new smokeless fire pits and outdoor electric heaters are also great season extenders.

Perla Sofía Curbelo-Santiago is a professional communicator, gardening promoter, and founder of www.AgroChic.com.

Just Add Some Water

Water can bring life-sustaining beauty to a garden. But you don't need to invest in an expensive pond or fountain to get its cooling effect: even the smallest water feature or a pebbly dry stream bed that suggests water can bring calm to your garden.

Place a birdbath in your perennial border or among low shrubs and plants that attract butterflies. Both birds and butterflies bring beauty and motion to a garden. Birdbaths range from basic bowls to eccentric art pieces that double as sculpture, but the best ones are shallow and roomy—about two to three inches deep and twenty-four to thirty-six inches across, with sides that slope gradually. Set it in a protected place, such as among trees or at the back of a border.

Create the illusion of water. If you live in a dry climate where water is precious, even the illusion of water can be soothing. For this desert garden, designer Troy Bankord added narrow channels of river rock that meander around and between circular pavers.

Meandering streams suggest tranquility. If you live where rain falls mainly in winter, try building a little dry stream bed. Dig a shallow trench, giving it a few curves. Nest some large boulders in the center, and smaller washed pebbles near the edges. Then, to visually soften and settle them into the landscape, plant low, mounding grasses along their edges.

If you have space for a swimming pool and would like to install one to cool off on warm summer days, consider a swimming pond instead of a regular pool, if local laws allow. These natural pools are created using plants and rocks as filtration for recirculating water. Half the pond's expanse is shallow, dedicated to water plants, and the other half is deeper, to allow for swimming. Do your homework regarding plants, because many aquatics are too vigorous for swimming ponds. If the description says the plant's spread is "indefinite," that won't be a good choice. Plenty of others are well-behaved. Without a heater and added chemicals, a natural pool offers the benefits of water to a wide array of wildlife—and you.

Add a birdbath. Even a small water feature can add tranquility to a garden. The plants that surround this one in a North Vancouver garden echo its caramel hues, especially the *Heuchera* 'Crème Brulee' in the foreground.

Install a "natural" pool. Even a flat stone containing natural depressions can serve as a water feature. Just fill those basins with water and watch how its surface reflects surrounding trees and clouds. This one on Vashon Island in Washington State edges a stumpery. Or dig a shallow basin in the soil and line it with heavy-duty, rubberized pond liner (available in rolls or sheets at hardware stores). For a natural look, edge it with stones, then fill with water.

Accent a bed. A favorite object or memento among your favorite plants will light up sweet memories every time you pass it. In this bed, which designer Davis Dalbok calls "The Sea Is Rising," a glazed ceramic hippocamps, a fish-tailed horse of Greek mythology, appears to prance through a cluster of succulents, including a variegated *Aeonium* (left) and a flame-hued *Euphorbia tirucalli.*

Plant for Fragrance, Motion, and Glow

Many plants are known for creating special effects in gardens. Some produce fragrant flowers that fill the air with seductive scents on warm summer nights. Gentle florals drift from angel wing jasmine and common heliotrope. Headier fragrances come from angel's trumpet, honeysuckle, or mock orange. Herbal scents, such as those from lavender and rosemary, spice the warm air on a sunny afternoon. Set all these near retreat areas, by a turn in a path, or beneath a window of your home, where their fragrance can waft indoors.

If possible, locate deciduous trees, such as flowering crabapples, eastern redbuds, and maples, where their foliage can be set aglow when backlit by a setting sun. Beneath them, plant shrubs such as smoke bush (*Cotinus*) or oakleaf hydrangea. Tuck in a few ornamental grasses that produce orange, red, or tawny-hued foliage or seed heads, which also glow when backlit. Examples include *Molinia caerulea* 'Mooreflamme' and *Carex testacea*. As an added bonus: Ornamental grasses with delicate, narrow leaves tend to shiver and dance in the slightest breeze.

Finally, place a few white-flowered plants in the fronts of borders. Unlike other colors that disappear into the shadows as night falls, white flowers stand out in the lingering twilight, or when touched by moonlight. Good choices for a summer border include common yarrow (*Achillea millifolium*); *Cosmos* 'Sonata White'; *Gaura lindheimeri*, whose delicate blooms appear to float on tall, wiry stems; and—for shady areas—white-flowered hydrangeas. Or fill a container with 'Casablanca' lily bulbs in spring, and you'll add both glow and fragrance to your patio in summer.

Whether it's a shapely ceramic jug that your mother gave you, a small totem you purchased during a trip to Alaska, or a polished wood birdhouse that a friend made for you in his garage, all can transform a quiet garden into a magical hideaway. They add a still moment, allowing your eye to rest and take in their forms before moving on.

Create Gathering Places

Some landscape professionals suggest that you include at least three places in your garden where people can gather—each with a different function, such as socializing, dining, or sitting by a firepit. Patios, either attached to the rear of your house or tucked beneath trees toward the back of your yard, are probably the most frequently used parts of your garden. If that spot will be primarily for entertaining, or for regular family meals in summer, locate it as close to your indoor kitchen as possible (unless you plan to add an outdoor kitchen, pizza oven, or portable barbecue on the patio as well).

Another solution to the small yard dilemma? Divide your space into two rooms, one for growing food and for dining; one for relaxing and entertaining. Let them flow together, as landscape designer Leslie Bennett has done in her small urban garden (page 170).

But if your deck or patio is small, consider flexible uses through the seasons. Put in a café table and chairs for late spring meals. Then when summer comes, remove the table and rearrange the chairs for informal "wine and cheese" get-togethers. And in fall, you might rearrange those same chairs again around a portable firepit for family gatherings. Whatever design you choose, light some lanterns and bring out the appetizers. Then sit back, relax, and watch the twilight deepen.

Opposite: **Small but smart.** Where space is limited, build in versatility. This small front yard, in Venice, California, is designed with family gatherings in mind. The front fence is all that separates this patio from the city sidewalk and the street, so it's designed to provide both privacy and light (thanks to its narrow windows). A lighted fountain trickles softly beside the front gate as it spills into a narrow rill leading into the patio—especially inviting to all who enter.

Make it cozy and add soft lighting. When night falls, lanterns glow softly and a gas-fed fire pit warms guests. (Not shown, but a fun idea: A projector behind the sofa projects movies onto a large pull-down screen.)

Leslie Bennett

GROW AND DISPLAY WHAT YOU LOVE—TIPS FOR MIXING FOOD PLANTS WITH ORNAMENTALS AND ACCESSORIES

Sitting at her computer at a law firm in London, Leslie Bennett made a decision that would ultimately change the direction of her life. "I was sitting at a desk," she recalls, "and it struck me: 'I feel disconnected from the land.'"

She headed to Jamaica to reconnect with the land and to her heritage (her mother is English; her father, Jamaican). There, she found work growing food at an organic farm while marveling at the beauty of fresh-from-the-soil crops. That experience brought meaning and a sense of purpose back into her life.

In Northern California with her husband and children, Leslie has finally reconnected with the land. She has filled her own garden with plants that express her rich heritage, and that are culturally meaningful to her family. They include lemon grass ("my dad's favorite, used in Jamaica to make tea"); passion fruit

**No room?
No problem.**
Although small,
Leslie's garden—
which she created
with help from
Designer Holly
Kuljian—includes
three outdoor living
areas. Spots for
lounging, dining, and
growing food are all
connected and flow
easily from one to
the other. Two raised
beds, each three
feet wide by ten feet
long, are filled with
tomatoes, herbs,
and cucumbers.

("It grows throughout Jamaica, and my son loves it"); and elderberry, which her mom once picked in her North Wales garden. "I want my kids to know where food comes from, and to tell them who we are."

Now working as a landscape designer, Leslie's focus is on creating ornamental-edible gardens. Her most important advice to clients now? "Make your garden truly reflective of who you are. Fill it with things that resonate with you at a soul level and make you feel at home."

Choose plants that are meaningful to you. Perhaps it's a lovely rhubarb plant that reminds you of your childhood home, those blueberries that you picked and found so refreshing while on a memorable family hike, or the lemon tree whose blossoms filled the air outside your window with sweet scents. Find a place for those plants that help you to recall sweet memories and that fill you with gratitude when you tend them.

Use fruit-bearing trees as ornamentals. Include types that bloom or bear fruit in different seasons, and arrange them for a seasonal show around your garden. In Leslie's garden, winter kumquats give way to apple and pluot blooms. Then comes quince and elderberry, followed by persimmon and pomegranate. Trees bursting into bloom or fruit at various times "make your garden more exciting," Leslie says.

Share your heritage. Like Leslie has done, seek out plants that remind you of places where other family members have lived. Heritage also comes from growing plants that lend their signature flavor to specific dishes of other countries, such as Thai basil, Cuban mint, or peppers for Hungarian paprika. Sometimes the seeds or starts are not in the general marketplace, but passed hand to hand and family to family. However, gardeners are generous, and often are willing to share seeds or cuttings that contribute to their own cultural cuisine.

Tuck ornamental edibles into your flower beds. Plant a big blue cabbage beside flowering thyme, for example. Or try growing edible flowers such as nasturtium or violas among herbs or vegetables. Edge flower beds with herbs—lime thyme in front of lavender, for example. Or plant creeping golden thyme in front of bluish culinary sage.

Make use of vertical surfaces. Plant grapevines near the base of a fence, for example. Or espalier apple trees against a simple post-and-wire trellis in a narrow side yard.

For more ideas, check out Leslie Bennett's book with coauthor Stefani Bittner, The Beautiful Edible Garden: Design a Stylish Outdoor Space Using Vegetables, Fruits, and Herbs. *Her website is www.pinehouseediblegardens.com.*

Go "natural." If you live near the edge of wild land, hike through it and note what grows there, naturally—whether in sun, or shade, on flat land or on hillsides. Then visit a native plant nursery in your area and seek advice on best plant choices. If possible, mimic those plant colonies in your garden—a process that author, nursery owner, and native plant specialist David Fross calls "wildscaping." His garden in Central California, pictured here, mimics a meadow filled with various grasses, including wild rye (***Leymus condensatus*** 'Canyon Prince') and flanked by native shrubs. He can stroll that gravel path beyond its bend, to savor its stillness and beauty at day's end.

HEALTHY PLANET

"The fun of gardening is to work with nature and to feel a part of it."

—Ciscoe Morris, author, garden expert, TV and radio personality

Make a Difference Through Gardening

How can you use your love of gardening to make a difference in the world? In Part 3, you will discover the ways people are joining together to build healthier ecosystems through restoration, regeneration, and reforestation. When home gardeners pool their knowledge and time to help make neighborhoods, communities, and wild lands better, everyone benefits.

You could focus on gardening practices that won't negatively affect the wild lands beyond your back fence. Or you may volunteer at a community garden, plant local street trees, or pull invasive weeds from wild lands. You might simply choose to support organizations that bring the natural world to the whole community—your nearby public garden or the Nature Conservancy. No matter how small you think your contribution might be, every bit counts. And by getting involved, you will discover the magic that happens when neighbors work together, learn a few new gardening skills, and make like-minded friends during the process.

No garden stands alone. All are part of the greater landscape. By mid-August, in Stan Fry's New Hampshire garden, red maples glow against dark green conifers, while coneflowers (*Echinacea*), phlox, bee balm (*Monarda*), and spikes of purple *Verbena bonariensis* mingle like floral confetti. They always attract visitors— pollinators, birds, and other wildlife from the forest lands beyond.

Meet your region's native plants. To discover some of the best planting ideas in your region, look no further than botanical gardens in your area. The one pictured here, Mt. Cuba Center, is nestled among the gently rolling hills of the Delaware Piedmont in Hockessin, Delaware. Here, you will discover an enormous collection of plants that are native to the eastern seaboard, along with a diverse set of planting communities, including a lilac allee, meadow, and woods garden—all surrounded by natural lands. In the pond garden, woodland groundcovers carpet the land.

Plan Your Garden for Positive Impact

By recognizing the connection between your privately owned habitat and the environment beyond it, you can begin to make any needed improvements. For example, if you don't have a rain garden, build one to prevent storm water runoff. Otherwise, excess water could flow off your property, then out over streets and other paved surfaces, where it could pick up pesticide and herbicide residues, excess nutrients, oil, and trash that can pollute lands and natural waterways. Just dig a wide, shallow depression in the soil near your home's downspouts to capture rainfall and slowly release it into the soil. Make it slightly deeper in the center where water collects, shallower toward the sides. Set plants that can tolerate wetter conditions in the center, and ones that prefer drier condition toward the edges. If possible, use porous paving on your property—such as crushed gravel, instead of solid concrete—so rainfall can pass through it instead of running off of your property.

Another effective strategy is to incorporate into your garden some of the plants that grow naturally nearby, and arrange them as they might occur on wild land. Hummingbird sages might find a home beneath oaks in California, for example, or bluebonnets among meadow grasses in Texas.

In his latest book, *Nature's Best Hope*, entomologist Doug Tallamy points out how your individual garden can play a major role in restoration. As he notes, 85 percent of land east of the Mississippi is privately owned. So, if everyone who tends a piece of land designated a place for planting natives on their individual plots, we could reconstruct healthy and interconnected ecosystems for birds and pollinators. The accumulated impact could create massive restoration.

Other ways to reduce your impact beyond the garden fence are mentioned throughout this book. Some of the most important include:

- Avoid using any plant designated as "weedy" or "invasive" in your area. Such plants can escape onto wild lands, where they will eventually crowd out native plants.

- Plant trees and shrubs, as they capture more carbon than herbaceous flowers.

- If you live near a stream, lake, or natural pond populated by salmon or other fish, leave any existing natives on the shoreline undisturbed to avoid polluting that waterway.

- Eliminate lawns to prevent runoff, or select drought tolerant grasses.

Teresa Speight

GROWING COMMUNITY THROUGH GARDENING

"Communities help you become rooted to the earth," says Teresa Speight. As a Master Gardener, Teresa was trained to help people get started with plants. Her goal was for "people to grow something, anything," she says. But when COVID-19 struck in 2020, Teresa realized that by teaching her neighbors to garden, she could also strengthen the ties in her community.

Teresa's home is in a Black middle- and working-class suburban neighborhood in central Prince George's County, Maryland.

In the spring of 2020, when the pandemic kept so many at home, food security was on the minds of many, Teresa says. Plots in community gardens filled, with long waiting lists for openings. So Teresa began to plant.

She put extra seedlings of all the warm-season vegetables—peppers, tomatoes, eggplants, squash (and flowers for fun)—out on her driveway with a sign that said "free." All disappeared before noon. She also anonymously delivered thirty-five plants, with detailed instructions, to doorsteps around her neighborhood.

Along with the plants, Teresa offered advice. She noticed that a gardener posted on a neighborhood app that her squash plants had gotten fungus and she'd pulled them all up. "I replanted a pot of squash for her," Teresa says. The two of them also discussed how far apart to plant next year's squash seedlings to allow for better air circulation.

Now the neighborhood is planning a seed and plant swap. They're continuing online monthly meetings. Others are following Teresa's lead and stepping up to share extra plants, tips, and expertise—from bonsai advice to suggestions for the best raised beds.

This interest in healthy gardening has changed Teresa's neighborhood for the better, because the community met its challenges in a positive way. "If we can do it in the bad times," she says, "then it will be easier to come together in the good times." If you're interested in doing this kind of garden organizing where you live, here are some of Teresa's tips.

Don't be shy about sharing. Most everyone is pleased to receive a gift. Seed packs often have more than a backyard gardener needs. Share the extras—either as seeds or as small starts, like Teresa did.

Draw beginners into gardening. Offer advice when it's asked for. And if you're a beginner, go ahead and ask questions of your more experienced neighbors.

Use online tools. Explore apps and platforms that can connect you to your neighbors. Through a neighborhood app, Teresa answered a request for seeds from a father who had built an indoor hydroponic garden because he lived in a townhouse. "I donated several types of seeds for him to enjoy and explore, from carrots to assorted lettuces," Teresa says. "He wanted to introduce his children to options in growing."

Recycle and reuse. Share used containers among those in your neighborhood. Someone else may have need of your stack of one-gallon pots if you let them know they're available. Encourage people to bring back the containers used for giveaway plants. If you have extra tools, go ahead and share them with nearby gardeners. If you set the example, others will follow.

Share costs. Equipment, like a good leaf shredder to make great mulch, is expensive. But if the cost of the purchase and upkeep is split among neighbors, owning an efficient machine becomes doable.

Use gardening to find common goals. For example, you might decide to get together with the neighbors and grow more food than you need, in order to give the extra harvest away to your food bank. Or you could choose to beautify your neighborhood with plants, as some communities have done by planting flowers in curb strips along their streets. As a group, you could pool your resources to buy trees at a lower cost. That's creating a healthy garden and growing a healthy, connected community.

Discover more of Teresa's interests and plenty of great garden information at her website, www.cottageinthecourt.com.

Opposite: **Growing relationships.** Community gardens are great places to meet other gardeners, learn some tips, or even swap crops. At Fulton Community Garden and Display Garden in Portland, Oregon, plot holders can grow produce in the ground or in raised beds. They also swap crops with fellow gardeners, or share the harvest with those who need it.

Community Gardens Bring Health to Neighborhoods—and the Land

Community gardens are open spaces, usually in urban settings, where people come together to plant, plan, and share everything from produce to education. Demand for these urban garden plots has been on the rise. Waiting lists to participate are long. The need for public places where people can dig in the dirt is driven by practical reasons—many people live in buildings without land. There's also the need for vegetables and fruits in neighborhoods that have limited access to healthful and affordable fresh food (aka food deserts).

Even if you already have a home garden, supporting community gardens in your neighborhood allows others to enjoy the health benefits that working with the soil gives. Here are ideas and examples of how people are organizing and benefiting from these growing spaces.

If your plants deliver more crops than you and your family can use, consider donating your excess fresh fruits and vegetables to those in the community who need them. At Fulton's Community Garden, those crops are gathered through the garden's "Produce for People" program.

Helping out can be as simple as tucking extra seeds in a waterproof baggie and tacking it to the gate of your community garden. Assistance doesn't even have to be garden related. In one corner of the community garden, you could organize a children's play group—for storytelling or outdoor art projects—giving parents that needed half hour to delve into their plots. To keep a community garden functioning, all kinds of skills are in demand, from grant writing to teaching healthy seed starting. Figure out what works for you.

An interesting twist on public garden projects comes out of Philadelphia's Thomas Jefferson University. Students in the landscape architecture program, using a grant from the Greenfield Foundation, have created Park in a Truck. Neighbors help decide on designated land and choose from a series of prototype garden designs, including an edible garden. All the components needed are loaded onto a truck, and students and neighbors work together to construct the garden. Though it is still in the early stages, the program sees success when local people buy into the process, making it a true grassroots initiative.

Or how about a pop-up community garden that has no plans to be permanent? That's the Peterson Garden Project on the north side of Chicago, founded in 2012 by LaManda Joy, who had the idea of using temporarily bare land for a minimum of two years and filling those lots with a series of raised beds. After that, if the ground is slated for development, the project moves on to another location. Some people have followed the pop-up gardens to three different places, because, as LaManda says, "Ten percent of a community garden is garden. Ninety percent is the community." Could this model work where you live?

On the other hand, Seattle's P-Patch program is an example of what can happen when a community garden just keeps on growing. Founded in the early 1970s, it now has eighty-eight gardens and two thousand plots on twelve acres, partnering with the city and other organizations to include market and youth gardening, and community food security programs.

Want to start a community garden? The P-Patch mission statement suggests what can be accomplished through gardening. Their goals include "improving access to local, organic, and culturally appropriate food; transforming the appearance and revitalizing the spirit of their neighborhoods; developing self-reliance and improving nutrition through education and hands-on experience; feeding the hungry; preserving heirloom flowers, herbs, and vegetables; building understanding between generations and cultures through gardening and cooking." You might begin with any one of these objectives and add more as you go.

Any time a city takes a greener approach in their land-use laws, the natural world benefits. In Atlanta, Georgia, the Food Well Alliance is in partnership with many organizations and people, from the Atlanta Botanical Garden to growers, community leaders, and volunteers, whose goal is to build a network of community gardens and urban farms. But most important for the restoration of a healthier bit of earth, Atlanta has also allocated resources to develop a city-wide planning process that will reserve precious open spaces specifically for future urban farming.

These ideas and actions are popping up all over the country. You can't usually work for change by yourself. But when like-minded people set out to make healthy gardening in public spaces a goal, it can happen.

A store entry welcomes shoppers with a garden. Farm on Ogden is a partnership of the Chicago Botanic Garden and Lawndale Christian Health Center, operated by Windy City Harvest. The store entrance is filled with a delightful mix of ornamentals and edibles grown together. Inside the massive building, greens are grown year-round under lights. In conjunction with the health center, prescriptions for fresh produce (Veggie Rx) can be filled at the farm store.

Abra Lee

PUBLIC GARDENS—TAKE ADVANTAGE OF GREAT IDEAS IN A TIME OF REINVENTION

If you've visited a public arboretum or botanic garden lately, you've strolled past displays that focus on local and native plants. But what you see in the gardens is only one part of a bigger picture. These organizations have always had a vital mission of education, conservation, and research. However, right now, many are making changes that will benefit you and, by extension, the living world around you. According to horticulturist and Longwood Fellow Abra Lee, these institutions are asking important questions about how they can reach out and best support their own communities. "There's an urgency," she says, "to connect people with outdoor spaces. I think botanical gardens are at the point where they can reignite, and replenish, and inspire, tapping into imagination to create places accessible to all."

Abra should know. She's a graduate of the prestigious Longwood Fellow leadership program, where she had an insider's view on how public gardens are shifting their programming, services, and approach. In order to ensure a bright future, they are

A forest in the city. Early-morning sunlight filters through the trees onto the Kendeda Canopy Walk, a raised boardwalk winding through a fine collection of southern hardwood trees in the Atlanta Botanical Garden. Abra Lee cites the organization as a good example of a public institution that is actively working to engage and welcome the diverse community they serve.

stepping up their efforts to be even more accessible and inclusive for all people, regardless of gender, race, socioeconomic status, or physical ability. She says, "Longwood showed me what was possible."

For example, many institutions are now focusing on grassroots partnerships beyond their gates—meeting people where they are, by sponsoring community gardens, neighborhood rehabilitation projects, and other community-based initiatives. And the more people who engage in healthy gardening, the better the chances for revitalization and restoration in the natural world.

If you have a public garden or arboretum nearby, now is the time to take advantage of the wealth of knowledge they offer. Or if you visit one during your travels, enjoy the delightful experience of exploring the gardens it contains. (For a sampling of great public gardens, see Resources on page 198.) By learning from and supporting arboretums and botanic gardens, you can help create a greener future for everyone. Here are Abra's tips for how to get the most out of what public gardens offer.

Discover great new (to you) plants. Arboretums and botanical gardens have a plant-purchasing budget far in excess of yours, but they don't waste money. Instead, they grow garden-healthy choices that work in that specific climate. And more places than ever are focusing on plant collections native to their region. Check plant labels and take a quick picture of the plant tag. You can research online later for more information.

This will help you build a solid wish list for what will work at home. Also, unlike in a nursery, you'll see plants grown to full size, which helps in your design planning.

Go for the education. Your local public garden can provide a learning experience through their ever-changing displays. It may offer speakers' presentations—either singly or as multi-day programs. Presenters are often engaging, as they provide useful gardening information and share the latest updates about plants and healthy plant practices that you can put into play in your own garden. The more you visit public gardens, the more takeaway ideas you will find. Money-saving memberships are frequently available.

Provide feedback. If you have ideas for improvements, Abra says, speak up. For instance, if people in your area of town have no access to fresh produce, you might enlist the help of your local botanic organization to start a community garden. As Abra learned during her Longwood program, institutional leaders are often willing to discuss innovations; in fact, the best ones are always on the lookout for good ideas from community members. As the needs of neighborhoods and the natural world continue to shift, these well-established institutions are evolving to meet the demands. And you can be part of making that positive growth happen.

For more about Abra Lee's wide-ranging work, which she says "ties together garden history, pop culture, and fashion," visit her website at www.conquerthesoil.com.

Support a Healthy Ecosystem with Your Food Dollars

When you visit your local farmers' market, you promote your own health and bolster the livelihoods of farmers who engage in sustainable methods. The fresh produce you buy might have been harvested that morning—almost as good as your backyard, in terms of travel. You can't beat the taste. And you'll be eating seasonally, another healthy thing to do.

You can also consider buying a box (or a half-box) from a community-supported agriculture (CSA) farm. These programs allow you to purchase a future share of the harvest, with weekly pickup or delivery boxes. Some have flexible shopping, where you get to choose which fruits or vegetables appear in your box that week. Many CSA farmers reach out to their customers, encouraging farm visits and offering recipes for produce. Building community relationships is an important aspect of belonging to a CSA.

Or you might connect with environmentally friendly farmers by shopping online. Farmers and ranchers are now joining together, so they can enlarge their markets and selling power. You'll find a wide range of organic produce and meats as well as fresh-caught seafood, delivered to your home within three days.

Support Street Trees for Your Neighborhood

Some people see street trees as just another maintenance headache. Deciduous types, especially, drop leaves that require raking and clog street gutters; they demand pruning, and they lift sidewalks. Other people, such as Dan Burden, head of Walkable Communities, a Florida organization that helps municipalities design and plan for tree success, say that the benefits of street trees far outweigh the disadvantages.

Dan says it's now well documented that trees have a calming effect on drivers, and on traffic. Trees mark a visual barrier between sidewalks and streets, making it safer for pedestrians and drivers to stay where they belong.

Brent Green

FIND A NEED AND FILL IT—GROW YOUR COMMUNITY BY PLANTING APPROPRIATE SHADE TREES

You might be thinking about planting one tree. Or maybe a couple. But Brent Green, a landscape designer in Los Angeles, took that idea one step further and changed his whole neighborhood.

Brent has loved trees since he was a boy growing up in the Leimert Park area of Los Angeles. There, he worked in his mother's garden and used his allowance money to buy plants at a local nursery. After earning his college degree in horticulture at Cal Poly, San Luis Obispo, he returned home to LA and began designing home gardens throughout the city.

But the lack of street trees where he and his wife settled in southeast LA bothered him. The nearby I-10 freeway created lots of smog, which caused local spikes in asthma and allergies. When he learned the city had no funds for street trees, Brent took it upon himself to plant them.

"I planted thirty-five trees along my neighborhood streets to celebrate my thirty-fifth birthday," he says. "I got planting help from my neighbors." One of them sent out an email, and the word spread; more volunteers came to help. Local nurseries began offering him discounts on trees that had grown too big for their containers or were otherwise not selling. Brent used his own money to buy the trees—and, come each birthday, he plants more. "I call them my birthday trees," he says. All are appropriate for the local climate and thrive without fuss.

Why care so much about trees? For one thing, trees help muffle the sound of traffic, Brent says. Another plus: Trees make streets cooler in summer. But the impact on your neighborhood can be even greater. "Plant the right tree in the right place and you really can change the environment," he says. "Crime in this area is down by about forty-five percent, and that drop has been attributed to the growing tree canopy."

Although Brent still runs a bustling landscape design business, he remains committed to getting street trees planted. "Give blighted areas some love," he says, "and everyone benefits." Here are Brent's tips:

Get permission to plant trees on city-owned land. Some cities have local forestry departments whose workers are responsible for planting, tending, and pruning street trees. "Our neighborhood did not have that, which is why we took it on," Brent says. "Planting trees was a game changer."

Teach new gardeners to make a difference.
Inspired by Brent's efforts, local schools
have started volunteer groups to help plant
shrubs and trees. Kids who volunteer learn
about the power of greenery to change urban
environments, and to make a big difference
in cities over time. It's "the zen of it," Brent
says, that touches people.

Choose the right trees. Obvious choices are
those that do not demand lots of pruning,
feeding, or watering once they've gotten
established. Ideally, choose compact kinds
that stay well below power lines and are
adapted to your location and climate. For
his neighborhood, Brent chose Chinese
pistache (*Pistacia chinensis*), *Koelreuteria*
(various species), olive (*Olea europaea*), and
water gum (*Tristaniopsis laurina*), among
others.

*To find out more about the landscapes
Brent designs, visit his website,
www.greenartlandscape.com.*

A good street tree must be able to grow in confined and compacted spaces without bringing its roots to the surface. It should not grow so tall that it entangles its limbs in the power lines. One example is *Zelkova serrata* 'Wireless'. As its name implies, it's been bred to spread.

If you get the chance to plant a street tree, do your research as if it were for your backyard. Check eventual height and spread. Many communities have lists of what can be planted, but often these lists do not reflect the latest thinking in urban landscaping. If you must choose from their list, make your decision based on information you look up. If a certain tree will eventually grow to a hundred feet, it might not be the best choice, even if your municipality suggests it. Or join a volunteer group in your community to help plant trees where needed.

Tree gender matters. Male trees produce pollen, and female trees catch it. Some trees produce little to no pollen. Notes Tom Ogren, author of *Allergy-Free Gardening*, allergies are on the rise because many towns plant male trees so they don't have to deal with the fruit from female trees.

What to do? If the existing trees where you live are heavy pollen producers, Tom advises planting female trees of the same species to help catch the drifting pollen before people breathe it in. You can also seek out trees with low pollen counts. To help inform gardeners, Tom has devised a system, the Ogren Plant Allergy Scale (OPALS), that rates the pollen in more than three thousand trees shrubs, flowers, and grasses. By paying attention to pollen when you choose trees, you can make the air in your neighborhood healthier.

Opposite: **Street trees mark the public and private spaces—beautifully.** The smooth trunks of a pin oak (left) and the rough bark of a pine (right) rise in a canopy over the sidewalk, yet allow glimpses of the home behind without the need for a privacy fence.

Walk with the trees. A graceful basswood tree (*Tilia americana*) makes a fine companion standing in the variegated hosta along a walkway in Chicago's Morton Arboretum.

191

Ron Vanderhoff

PROTECTING NATIVE PLANTS AND THEIR HABITATS

During the week, Ron Vanderhoff is more than busy with his day job as general manager of Roger's Gardens, a destination nursery and garden center in Corona del Mar, California. But come weekends, this native plant expert and conservation ecologist heads for the trails that skirt the nearby seaside, canyons, and mountains to search for and document native flora. Among his concerns: Invasive plants are crowding out native flora that could otherwise benefit the region's ecosystem; native habitats need protection from invasive plants that have escaped from home gardens; and the habitats of migrating Monarch butterflies are disappearing, along with the milkweeds upon which they feed.

Ron leads plant-related field trips into natural areas each year. On these trips, he highlights problems created by non-native invasive species. "I will pause the group for a few minutes on a pretty hillside that's overrun by invasive plants that do not belong here—like pepper trees or tamarisks," he says. "I will ask the group to listen as they look over the scene. All is still—no motion, no birdsong, nothing." Then Ron will move the group to a similar-looking area, dominated by native plants. "And, sure enough," he says, "the air is alive with the sounds of insects buzzing, birds chirping, lizards scuttling among the bushes, butterflies flitting among blossoms. This habitat is alive!"

When visiting natural areas, look and listen carefully in order to understand if an ecosystem is functioning as it should. Why should you care? Non-native, invasive plants do not support other beneficial organisms, because they did not evolve alongside them. Also, plants that invade natural areas negatively impact historic fire regimes, soil erosion, groundwater pollution, and human recreation. "Together, we can reduce the invasive plant problem dramatically if we first educate ourselves about its significance, and then follow up with our shopping dollars and our voices," Ron says. Here are his tips:

Get to know your region's true natives. "Horticultural invasives," sometimes sold as landscape plants at garden centers, don't stay put in gardens. Depending on where you live, certain kinds can creep into wild lands, where they establish and multiply. There, they can outcompete native flora, because they have few natural predators or environmental controls to keep them in check. Examples include English ivy, privets, running bamboos, and periwinkle. Check local online sources to find the best garden-friendly natives in your region.

Shop wisely. If possible, seek out native plant specialty nurseries in your area. Keep in mind that a plant that's invasive in one region of the country may not be in another. You may need to do some additional research, because, as Ron notes, the "invasive plants that you might find in a garden center do not have flashing red lights on them." However, you don't have to be a native plant purist, says Ron—mix and match them with your cultivars. He says, "Don't forget native milkweed. Monarch butterflies depend upon this plant, and gardeners can play an important role in their conservation and recovery."

Volunteer to help clean up wild lands near you. Seek out your state's stewardship organizations; they almost all have great volunteer programs for weed eradication and more. For instance, in Ron's part of California he references the Irvine Ranch Conservancy, the Laguna Canyon Foundation, and the Newport Bay Conservancy. In addition, most of the larger public/private land trusts around the US also offer volunteer opportunities.

Ron Vanderhoff is director of PlantRight, an invasive plant council, and contributor to various books, including Wildflowers of Orange County *and* The Sunset Western Garden Book, *as well as the California Native Plant Society's website.*

Help to steward wild lands. Throughout the country, wild lands are staying healthy and vibrant, thanks to help and support from volunteers. Here, a trail guide snaps a photo at Caines Head State Recreation Area above Resurrection Bay, Alaska.

Help to Protect and Restore Wild Lands

If you are interested in the natural world, chances are that you are keenly aware of environmental changes now impacting plant and wildlife communities in your area and beyond. As urban sprawl continues to creep, natural habitats everywhere are shrinking.

Get acquainted with the issues facing parklands and wilderness areas near you, or those that you have visited and care about. In the Pacific Northwest, you can work with the US Forest Service to act as stewards for the wilderness, by repairing trails and removing invasive plants. In Michigan, where grassland conservation is gaining in importance, many projects are aimed at restoring acres of diverse native prairie grasses and establishing new native grassland habitats for birds, pollinators, and small mammals. In Louisiana, where coastal wetlands are vanishing into the Gulf of Mexico at an alarming rate, due to natural causes and human intervention, you can join other volunteers who are working together to help grow and replant native trees and grasses. In Ohio and Vermont and in many other states, volunteers are working together, through stewardship programs sponsored by the Nature Conservancy, to help remove invasive plants from natural lands. And in Hawaii, volunteers from across the state are working to restore native tree populations that have long since disappeared from the volcanic slopes by pulling invasive weeds.

As your time permits, you can join other volunteers working to preserve public lands. By helping to restore and maintain natural marshes, grasslands, and forests impacted by human activity, you will contribute to the health of this planet we call home. (For a list of volunteer opportunities across the US, see Resources on page 198).

Look to the Future: Leaving a Legacy of Health

You accumulate knowledge on healthy gardening one small step at a time. Through your experiences—both successes and failures—you gather practical information, as well as an appreciation for the interconnection of all that lives in the world. And the last step in the cycle is to pass on what you've learned to your friends, neighbors, and the next generation of gardeners.

Most of the experts in this book hold childhood memories of gardening or of being in nature. Just as we encourage toddlers' good eating habits by offering them a wide range of tastes in their first foods, so we can also start early to engage children in the vast variety of outdoor pleasures. Share your enthusiasm and allow them to experience nature's magic in a positive way. Give them hope to grow on. Later they can learn about the earth's environmental woes. You love something first; then you want to save it.

But passing on a legacy of healthful gardening doesn't stop with children. Reach out to others in community center horticultural programs or retirement homes. Join a gardening group, and then welcome diversity in an ever-widening circle of new members.

And keep on learning. Nature always has more to teach us. After all, *you* can't go stand outside and make lunch for yourself from the sun's rays—but every plant can. So, a certain humble perspective about gardening may be the greatest gift you leave behind.

By joining with others, you have the power to create change. Each small step makes your garden and your community more sustainable. It's like sowing the tiny seeds that will someday grow into mighty forests you won't see—but they're worth planting anyway. Who knows? The creative and problem-solving innovations of future generations may already be germinating, hidden in the earth, waiting for the right time to bloom.

"When you tend a garden, thoughtfully, with respect for all living things in it, everyone benefits, from the smallest insects to the tallest trees."

—Kathleen and Mary-Kate

Resources

Chicago's multicolored meadow. Bands of blooms grace Lurie Garden, which sits atop a parking lot. For this summer show, garden designer Piet Oudolf selected plants that come back year after year without much tending. They include blue *Salvia × sylvestris* 'Rugen', commonly called wood sage (foreground), and violet blue *Salvia nemorosa* 'May Night' (behind).

Our Favorite Public Gardens: A Sampling

If you're new to gardening or are seeking fresh planting ideas, head for a botanical garden or arboretum near you. To find additional botanical gardens, visit www.public gardens.org.

Washington Park Arboretum, Seattle, WA. Plants from western Washington fill this two-hundred-acre garden on the University of Washington campus. *botanicgardens.uw.edu*

Idaho Botanical Garden, Boise, ID. These 50 acres feature specialty gardens that highlight natives of the intermountain region, plus fire-wise and water-wise plants. *idahobotanicalgarden.org*

Arizona-Sonora Desert Museum, Tucson, AZ. Plants of The Sonoran Desert Ecosystem are featured here. Don't miss the Pollinator Garden or the Desert Demonstration Garden. *www.desertmuseum.org*

Lady Bird Johnson Wildflower Center, Austin, TX. More than nine hundred species of native Texas plants grow throughout these 284 acres, including cultivated gardens, managed natural areas, water features, and gardens for dry creek beds. *www.wildflower.org*

Denver Botanic Gardens, Denver, CO. This garden specializes in plants that are best suited to its mile-high location, including alpine and rock garden plants. Check out the use of stone troughs—easy to duplicate at home. *www.botanicgardens.org*

Brooklyn Botanic Garden, Brooklyn, NY. The fifty-two-acre garden is home to native flora, woodland plants, and more. *www.bbg.org*

Missouri Botanical Garden, St. Louis, MO. Trees and meadows fill this seventy-nine-acre garden. Visit the garden's website to learn what's in bloom and how to build sustainability into your own garden. Or take a virtual garden tour. *www.missouribotanicalgarden.org*

North Carolina Botanical Garden, Chapel Hill, NC. Native plants are the focus of this garden, on 1,100 acres of conservation lands, including grasslands and bogs. Don't miss the wooded Piedmont Nature Trails behind the display gardens. *ncbg.unc.edu*

Fairchild Tropical Botanic Garden, Coral Gables, FL. Tropical plants fill these eighty-three acres of lowlands, lakes, and palms. *fairchildgarden.org*

Volunteer Resources

Join other gardener-volunteers to help improve wild lands and communities.

Find a cause or a place that you care about, whether it's tending imperiled forests, deserts, prairies, wetlands, swamps, or coastal ecosystems. Then sign on to help. Find details at the Bureau of Land Management website (blm.gov/get-involved/volunteers); the National Park Service (nps.gov); the US Department of the Interior (doi.gov; National Public Lands Day); and Hawaii Forest and Trail Hawaii (www.hawaii-forest.com).

PLANT STREET TREES

To start, visit the Arbor Day Foundation website (arborday.org). A sampling of established volunteer groups is listed here.

California: California ReLeaf, californiareleaf.org

Kentucky: Trees Lexington!, treeslexington.org

Louisiana: Common Ground Relief, commongroundrelief.org/plant-a-million-trees

Minnesota: Tree Trust, treetrust.org

New York: MillionTreesNYC, milliontreesnyc.org

Ohio: Trees Columbus, treescolumbus.org/projects

Oregon: Friends of Trees, friendsoftrees.org

Pennsylvania: Pennsylvania Horticultural Society, phsonline.org/programs/tree-programs

Tennessee: Nashville Tree Foundation, nashvilletreefoundation.org.tree-planting-events

Texas: Texas Trees Foundation, texastrees.org/volunteer

HELP PROTECT WILD LANDS

Depending on where you live, you can volunteer to help tend imperiled beaches or swamps, prairies, deserts, or forests. A sampling:

Surfrider Foundation. Coastal and ocean preservation is the focus of this international organization. Among its missions: Restoring beaches that are shrinking due to natural erosion and development. *surfrider.org*

Common Ground Relief. Louisiana's coastal wetlands, which provide about 280,000 acres of National Wildlife Habitat, are disappearing at an alarming rate; every year, some 10,000 acres of land vanishes into open waters. Volunteers help to replant swamp trees and grasses that can help stabilize this vital green-scape. *www.commongroundrelief.org/wetlands*

Mojave Desert Land Trust (MDLT). Protecting plants and wildlife of the California desert from North of Death Valley National Park to the Mexico border, is the goal of MDLT. Trained Land Steward volunteers help to habitat restoration, trail work, and planting—using materials grown in the organization's Native Plant Restoration Nursery. *mdlt.org*

Herbert Hoover National Historic Site. Located in West Branch, Iowa, this is one of only a few protected prairies that remain in North America. Volunteers help tend this vital ecosystem by removing shrubs or collecting native plant seeds from the park's 81 acres of tallgrass prairie. *nationalparktraveler.org*

USDA Forest Service. Forests cover a third of America. They store and filter water, provide habitat for plant-and animal species, and sequester some fifteen percent of fossil fuel emissions. But they face challenges, from rampant exotic weeds to wildfires and introduced pathogens. Volunteers work with the Forest Service to hike the trails and act as stewards, repairing trails and removing non-native invasive plants. *www.fs.usda.gov-working-with-us/volunteers*

Lifelong Learning

The following resources will help you on your gardening way.

Further reading:

HEALTHY GARDEN ELEMENTS

Design. Find information on how to design a garden that works with nature.

Fearless Gardening: Be Bold, Break the Rules, and Grow What You Love, by Loree Bohl. Timber Press, 2021.

Floratopia: 110 Flower Garden Ideas for Your Yard, Patio, or Balcony, by Jan Johnsen. Countryman Press, 2021.

New Naturalism: Designing and Planting a Resilient, Ecologically Vibrant Home Garden, by Kelly D. Norris. Cool Springs Press, 2021.

Planting in a Post-Wild World: Designing Plant Communities for Resilient Landscapes, by Thomas Rainer and Claudia West. Timber Press, 2015.

Soil and Water. Keep soil healthy and water practices sustainable.

Growing a Revolution: Bringing Our Soil Back to Life, by David R. Montgomery. W. W. Norton, 2017.

Soil Science for Gardeners: Working with Nature to Build Soil Health, by Robert Pavlis. New Society Publishers, 2020.

Hot Color, Dry Garden: Inspiring Designs and Vibrant Plants for the Waterwise Gardener, by Nan Sterman. Timber Press, 2018.

Plants. These will help you research the most successful plants for your location.

Lawn Gone! Low-Maintenance, Sustainable, Attractive Alternatives for Your Yard, by Pam Penick. Ten Speed Press, 2013.

Shrubs and Hedges: Discover, Grow, and Care for the World's Most Popular Plants, by Eva Monheim. Cool Springs Press, 2020.

The Tree Book: Superior Selections for Landscapes, Streetscapes, and Gardens, by Michael A. Dirr and Keith S. Warren. Timber Press, 2019.

HEALTHY YOU

Your garden can sustain you—body and soul.

Edibles. This information will help make your harvests the best ever.

The Chinese Kitchen Garden: Growing Techniques and Family Recipes from a Classic Cuisine, by Wendy Kiang-Spray. Timber Press, 2017.

Complete Container Herb Gardening: Design and Grow Beautiful, Bountiful Herb-Filled Pots, by Sue Goetz. Cool Springs Press, 2020.

Gardening with Grains: Bring the Versatile Beauty of Grains to Your Edible Landscape, by Brie Arthur. St. Lynn's Press, 2019.

Grow Your Own Mini Fruit Garden: Planting and Tending Small Fruit Trees and Berries in Gardens and Containers, by Christy Wilhelmi. Cool Springs Press, 2021.

Lush Life: Food & Drinks from the Garden, by Valerie Rice. Prospect Park Books, 2021.

Tips and Tricks for Success. Raise crops undercover and foil pests.

Growing Under Cover: Techniques for a More Productive, Weather-Resistant, Pest-Free Vegetable Garden, by Niki Jabbour. Storey Publishing, 2020.

Ultimate Guide to Indoor Gardening: Grow Veggies, Herbs, Sprouts and More at Home, by Kim Roman. Fox Chapel Publishing, 2021.

The Vegetable Garden Pest Handbook: Identify and Solve Common Pest Problems on Edible Plants, by Susan Mulvihill. Cool Springs Press, 2021.

Gifts from the Garden. The many rewards of growing plants.

Natural Beauty from the Garden: More Than 200 Do-It-Yourself Beauty Recipes & Garden Ideas, 2nd ed., by Janice Cox. Ogden Publications, 2018.

The Scentual Garden: Exploring the World of Botanical Fragrance, by Ken Druse and Ellen Hoverkamp. Abrams, 2019.

The Wellness Garden: Grow, Eat, and Walk Your Way to Better Health, by Shawna Coronado. Cool Springs Press, 2017.

A Woman's Garden: Grow Beautiful Plants and Make Useful Things, by Tanya Anderson. Cool Springs Press, 2021.

HEALTHY PLANET

These books explain the vital connection between earth-friendly garden practices and the health of the natural world.

Over the Back Fence. Here's how to bring about positive change.

Garden Revolution: How Our Landscapes Can Be a Source of Environmental Change, by Larry Weaner and Thomas Christopher. Timber Press, 2016.

Nature's Best Hope: A New Approach to Conservation That Starts in Your Yard, by Douglas W. Tallamy. Timber Press, 2019.

A New Garden Ethic: Cultivating Defiant Compassion for an Uncertain Future, by Benjamin Vogt. New Society Publishers, 2017.

Attract and Maintain Wildlife. Discover why home gardens are vital.

Garden Allies: Discover the Many Ways Insects, Birds & Other Animals Keep Your Garden Beautiful & Thriving, by Frédérique Lavoipierre. Timber Press, 2021.

The Humane Gardener: Nurturing a Backyard Habitat for Wildlife, by Nancy Lawson. Princeton Architectural Press, 2017.

The Monarch: Saving Our Most-Loved Butterfly, by Kylee Baumle. St. Lynn's Press, 2017.

Leave a Legacy. Here are books to inspire the next generation.

Gardening with Emma: Grow and Have Fun, by Emma Biggs and Steven Biggs. Storey Publishing, 2019.

Harlem Grown: How One Big Idea Transformed a Neighborhood, by Tony Hillery, illustr. Jessie Hartland. Simon & Schuster/Paula Wiseman Books, 2020.

Michelle's Garden: How the First Lady Planted Seeds of Change, by Sharee Miller. Little Brown Books for Young Readers, 2021.

COACHING, CLASSES, PODCASTS, AND WEBINARS

Look for gardening teachers who have a track record, such as Ron Finley at MasterClass, Melinda Myers at The Great Courses, and Theresa Loe's Canning Academy. To find a garden coach near you, the website Gardenary.com lists 180 garden coaches across the country. Follow blogs like *A Way to Garden* with Margaret Roach, *Cultivating Place* with Jennifer Jewell, and the many writers at *Garden Rant*. You can join groups on Facebook that range in erudition from The Garden Professors to The Beginner's Garden, as well as those many groups focusing on specific plants. If you enjoy listening to podcasts, you could try *Let's Argue About Plants*, from *Fine Gardening* magazine; *Plantrama—Science, Art & Dinner* with Ellen Zachos and C. L. Fornari; *The Native Plant Podcast* with John C. Magee; *Cottage in the Court* with Teresa Speight; and the ever-popular *Joe Gardener Show* with Joe Lamp'l. These and much more audio gardening advice can be found at the streaming website tunein.com/radio/Stream-Gardening-g229/.

Acknowledgments

We wish to thank the gardening experts who so generously shared with us their time, tips, and insights for the Gathering of Gardeners, as well as many others whose advice appears throughout this book. We are also indebted to Dr. Tanya Hudson and other health care professionals who helped us to further understand the connection between gardening and health—both mental and physical—including Florence Williams, author of *The Nature Fix*, and Dr. Richard Taylor, head of the University of Oregon Physics Department. We are grateful to the many photographers whose gorgeous images grace these pages: Caitlin Atkinson, Brie Arthur, Rob Cardillo, Jennifer Cheung, Robin Cushman, Darcy Daniels, Frances Freyberg, Tom Grey, Caroline Greyshock, Sara Hall, Saxon Holt, Heidi Hornberger, Gemma Hart Ingalls, Bonnie Kittle, Jeff Lafrenz, Jason Lewis, Melinda Myers, Matt Thomas, Mark Turner, John Ulman, Rachel Weill, and Doreen Wynja. Finally, special thanks to photo stylist Linda Lamb Peters and her assistant, Mikala Rene, for their magical touches to some images in Part One.

A bird in the hand. Healthy gardening gives you a connection with nature in profound and unexpected ways.

Index

A

aeonium, *44, 46, 47*
agapanthus, 95, *95*
agave, *16, 22,* 43, *43*
Alaska, 194, *194*
allergies, 188, 190
Allergy-Free Gardening (Ogren), 190
amendments, 88, 89, 92–93, 109, 134
American chestnut tree, 66, *66*
American plum, 57
angel's trumpets, 93, *93,* 94
arboretums, 184–86, *185*
Arthur, Brie, *112,* 112–14, *113*
arugula, 114, 134
Atlanta Botanical Garden, 182–83, 185, *185*

B

Baldwin, Debra Lee, *42,* 42–44, *43, 44, 46, 47*
Bankord, Troy, 163, *163*
baobab tree, 22, *23*
Barker, Jennifer, *124–25,* 125
Barker, Lance, *124–25,* 125
barrel cactus, *16,* 22
basswood tree, 191, *191*
beds, 166, *166*
    herb, 141, *141*
    protection of, *110,* 111
    raised, 126, *126, 136–37,* 137
bee balm, 175, *175*
bees, 64, 126, *127–28*
begonias, 80, *81,* 95, *95*
benefits, health, 98, 102–3, *103*
Bennett, Leslie, *170,* 170–72, *171*
berries, 139, *139*
biodiversity, 45–51, *46, 48–49*
birdbath, 163, 164, *164*
birdhouse, 106, *106–7*
birds, 73, *73,* 163

Bloom, Jessi, 98, 129
blooms, 126, *127–28*
blue chalk stick, 43, *43*
blue fescue, *70–71,* 71
blue salvias, *70–71,* 71
borders, 68–69, 84, *84*
Brenzel, Kathleen Norris, 197, *197*
brown-leafed smoke bush, 66, *66*
Brunnera, 63, *63*
Buehler Enabling Garden, 160, *160, 161*
Buley, Nancy, *54,* 54–56, *55*
Burbank, Luther, 50
Burden, Dan, 187
butterfly weed, 77–78

C

Caines Head State Recreation Area, 194, *194*
calibrachoa, 95, *95*
California, *14,* 24, 42–44, *43,* 173, *173*
    Northern, *70–71,* 71, 108, *108,* 170–72, *171*
    Southern, *13,* 14, 41, 68–69, 133, *133,* 133–34, 135, *135, 146,* 147, 168, *168,* 188–89, *189,* 192–93, *193*
California fan palm, *13,* 14
California Native Plant Society, 193
calorie-burning, 102–3, 145, *145*
Canadian blasts, 93–94
cardoon, 66, *66*
Caryopteris 'Sunshine Blue,' *158,* 159
Cascade Mountains, 122–23
Charleston, 80, *80*
Chicago, 77, *77,* 182, 191, *191*
Chicago Botanic Garden, 160, *160,* 161, *161,* 162, 183, *183*
chicken coop, 106, *106–7*
China, *52,* 53

Chinese windmill palm, 66, *66*
cholesterol, 104
cisterns, 106, *106–7*
city, 185, *185*
climate, 41, 44, 133–34, 163, *163*
coastal daisy tree, 95, *95*
coffeeberry, 14, *14*
color, 25, *25,* 69
Colorado, 24, 28, *28–29*
comfort, *149,* 149–50, 162
community, 10, 175, 179–80, 186
community garden, 178–83, *179, 181, 183,* 186
community-supported agriculture farm (CSA), 187
compost, 17, 18, 85, 88
    amendments and, for soil, 134
    making, 89, 91
coneflower, 175, *175*
conifer, *52,* 53, *66,* 66–67, 175, *175*
Connecticut, *26–27,* 27
container, 56, 91, 113, *113*
    colorful, 79–84, *80, 81–83, 84*
    plants, 93, *93*
    vegetable grown in, 108, *108*
cool-season, 118, 121
cradles, *52,* 53
Creasy, Rosalind, 108, *108*
Creating Sanctuary (Bloom), 129
creativity, 103, 134
creeping Jenny, 80, *81*
crops, 115, *115,* 121, 123, 133–34
CSA. *See* community-supported agriculture farm
Curbelo-Santiago, Perla Sofía, *160,* 160–62, *161*

D

Dalbok, Davis, 166, *166*
Daniels, Darcy, 30, *31,* 32
date palm, 22, *23*
Delaware Piedmont, 176, *176*

Diboll, Neil, 50
Dichondra argentea 'Silver Falls,'
    93, *93*
disinfectant, 90, *90*
diversity, 51, 60
dock, 95, *95*
dogwoods, *82–83*, 83
Douglas fir tree, 50
drinks, 135, *135*
dwarf rosemary, *70–71*, 71

E
echinacea, 95, *95*
ecosystems, 177, 187
Emerald Spreader, 57
Ernst, Linda, *158*, 159
espalier apple trees, 172
evergreen conifer, 30, *31*
exercise, 144–47, *145*, *146*
experiments, zonal, 93–94

F
fairy slipper orchid, 50
false blue indigo, 77
fan palms, *13*, 14
Farm on Ogden, 183, *183*
ferns, *58*, 72
fertilizers, 17, 41, 88
Florida, 22, *23*, 24
flowering maple, 93, *93*
Foertsch, Tim, 40, *40*
food, 104–7, *105*, *106–7*
The Foodscape Revolution (Arthur),
    114
foodscaping, 112–14, *113*
Food Well Alliance, 182–83
Fornari, C. L., 24
four o'clock flowers, 95, *95*
Fross, David, 173, *173*
frost protection, 123
Fry, Stan, 175, *175*
Fulton Community Garden and
    Display Garden, 180, 181, *181*

G
garden, 18, 21, 24, 45, 85, 177. *See
    also specific types*
    color in, 25, *25*
    in community, 10, 175
    container, 95, *95*
    creativity enhanced by, 103
    forever *vs.* for now, 30, *31*, 32
    health benefits of, 98, 102–3,
        103, 144–47, *145*, *146*
    health of, 17, 19, 96–101, *97*, *99*,
        100–101
    landscape connected to, 175,
        175, *176*
    path in, 156, *156*, *158*, 159
    planning for, 33–38, *34*, *36–37*
    prairie, 28, *28–29*
    sustainability in, 10
    tools for, 90, *90*, 144
    vegetable, 104–11, *105*, *106–7*,
        108, *110*, *111*
    waste from, 89, 91
    water and, 92–95, *93*, *94*, *95*,
        163, *163*, 163–65, *164*, *165*
    well-being and, 10
gardening, 148–50, *149*, 162, 174
    community through, 178–83,
        179, *181*, *183*
    positive impact of, 175–76,
        175–76
    for tranquility, 151–53, *152*,
        153
gathering places, *168*, 168–69, *169*
Gattone, Toni, 147, *148*, 148–50,
    149
gloriosa daisy, 25, *25*
golden bamboo, 95, *95*
golden barrel cactus, 43, *43*
Gordon, Sylvia, 24
grasses, 67, 167
Green, Brent, 188–89, *189*
green aeonium, 43, *43*
Greenfield Foundation, 182
Grivas, Erica Browne, 102
groundcovers, 72–74, *73*, 159, 176,
    176
grow bags, 91, 114
Growing a Greener World, 86
Grow Planet (Haupt), 12

H
hammock, 156, *157*
harvest, 129–32, *131*, 132, *132*
    short-season strategies for,
        122–23
    weather influencing, 143
Haupt, Lyanda Lynn, 12
Hawaii, 135, *135*, 154, *155*, 195
health, 10, 196
    benefits of community garden,
        181–83, *183*
    of garden, 17, 19, 96–101, *97*, *99*,
        100–101
heliotrope, 80, *81*
herbs, 104, 129, 141, *141*, 142, *142*
heritage, 170–72, *171*
Himalayan blackberry, 50
Himalayan cobra lily, 63, *63*
Hippocrates, 104
Hockessin, Delaware, 176, *176*
horticultural therapy (HT), *160*,
    160–62, *161*
horticulture, 112–14, *113*, 184–86,
    185, 188–89, *189*
    experts in, 11
    skills of, 19
    therapy, *160*, 160–62, *161*
hosta, 58
houseboat, 94, *94*
HT. *See* horticultural therapy
Hudson, Tanya, 103
Hume, Ed, 121
hybrid geranium, 93, *93*
hydrangea, 30, *31*, 59, 73, *73*

I
Indiana, 102
International Society of Arboricul-
    ture, 55
irrigation, 96, *97*, 109
Italian shell bean, 94

J
Jamaica, 170, 172
Japanese forest grass, 30, *31*, *58*,
    66, *66*

Japanese gold cedar, 66, *66*
Japanese maple, 30, *31*, 56
Japanese snowbell, 55, *55*
Japanese umbrella pine, 57
Jett, Susanne, 41
J. Frank Schmidt & Son Co., 54
Joy, LaManda, 182
Jungles, Raymond, 22, *23*

K

kangaroo apple, 95
kangaroo paws, 25, *25*
Kendeda Canopy Walk, 185, *185*
kiss-me-over-the-garden-gate, 95
Kuljian, Holly, 170–72, *171*

L

Lake Union, 94, *94*
Lamp'l, Joe, *86*, 86–88, *87*
land, public, 188, 195
landscape, 45
    garden connected to, 175, *175*,
        176
    sustainability of, 18
    water conservation and, 41
lantana, *39*
Larry Weaner Associates, *26–27*,
    27
lawn, 38–41, *39–40*, 74
Lawndale Christian Health Center,
    183, *183*
leaves, shredded, 87, *87*, 88
Lee, Abra, *184*, 184–86, *185*
Leimart Park, 188–89, *189*
Lifelong Gardener (Gattone), 150
light, 108–9, 120, 169, *169*
Lilly, Bob, *92*, 92–95, *93*, *94*, *95*
limitations, physical, 148–50, *149*
long-blooming annuals, 32
Longwood Fellow leadership pro-
    gram, 184–86, *185*
Louisiana, 195
Lumsden, Vanca, 102
Lurie Garden, 77, *77*
Lush Life (Rice), 134

M

Mackey, Mary-Kate, 197, *197*
manzanita, 14, *14*
Massachusetts, 24
Master Gardeners, 55
mattress, 89, 91
May apple, 103, *103*
meadow, 40, *40*
'Meerlo' lavender, *70–71*, 71
Merciari, Sherry, 25, *25*
Merciari Designs, 25, *25*
Mexican fan palm, *13*, 14
Mexican feather grass, 67
Michel, Carol, 102
microbiota, 45
microclover, 74
micro-organisms, 85–89, *87*
milkweed, 77
Mississippi River, 177
mockingbird, 11, *11*
monarda, *26–27*, 27
monoculture, 41
Morris, Ciscoe, 174
Morton Arboretum, 191, *191*
moss, 103, *103*
Mt. Cuba Center, 176, *176*
Mt. Hood, Oregon, 54
Muir, John, 10
mulch, 87, *87*, 88, 89, 106, *106–7*
Munts, Pat, 122, *122*
Myers, Melinda, *76*, 76–78, *77*

N

nasturtiums, 129
natural materials, *52*, 53
Nature Conservancy, 175, 195
Nature's Best Hope (Tallamy), 177
New Hampshire, 175, *175*
New Mexico, 96, *97*
Niedermyer, Stephanie, 102
*The Northwest Gardener's Hand-
    book* (Munts), 123
Northwest Garden Nursery, 60–61,
    62–63, 63
Northwind switch grass, 67

O

oak tree, 14, *14*
O'Byrne, Ernie, *60*, 60–61, *62–63*,
    63
O'Byrne, Marietta, *60*, 60–61,
    62–63, 63
Ogden, Lauren Springer, 28, *28–29*
Ogren, Tom, 190
Ogren Plant Allergy Scale
    (OPALS), 190
orangebark stewartia, *31*, 32
orchids, *82–83*, 83
Oregon, 30, *31*, 32, 40, *40*, 54,
    82–83, 83, 102, *124–25*, 125,
    158, 159, 180, *181*

P

Pacific Northwest, 24, 50, *58*, 103,
    103, 195
pain, chronic, 148–50, *149*
pale purple coneflower, 78
paper reed papyrus, 73, *73*
path, 156, *156*, *158*, 159, 173, *173*
patios, 159, *168*, 168–69, *169*
Pavlis, Robert, 130
penstemon, 51
Penstemon pinifolius, 47, *48–49*
peonies, 78
peppers, 114, 130
Perennial Plant Association, 76
perennials, 28, *28–29*, 32, *58*, 76,
    76–78, *77*
pests, 85, 126, *126*
Peterson Garden Project, 182
Philadelphia, Pennsylvania, 182
phlox, 175, *175*
pineapple guava, 135, *135*
planning, for garden, 33–38, *34*,
    36–37
plant communities, native, 17, 24,
    42–44, *43*, 51, 173, *173*,
    192–93
    in Hockessin, 176, *176*
Plant Development Services Inc.,
    68–69

planting, of seeds, 118–20, *119, 120*
The Plant Lady, 114
plants, 45, 51, 60, 61, 167
    container, 93, *93*
    cut-and-come-again, 130
    discovery of, 186
    invasive, 192–93
    needs of, 123
    For Now, 32
    pollinator, 126, *127–28*
    recycling of, 18
    roots of, 134
    space for, 32, 69
Platinum Beauty lomandra, *70–71,*
    71
pokeweed, 95, *95*
pollen, 190
pollinator season, 64
pool, 165, *165*
potatoes, 114
potted hair sedge, *62, 63, 63*
prairie dropseed, 78
prairie garden, 28, *28–29*
Prairie Nursery, 50
Prince George's County, Maryland,
    178–80, *179*
public garden, 184–86, *185*
purple coneflower, 25, *25, 26–27,*
    27

R
rainfall, 96, *97,* 109, 177
ramada, Southwest-style, 155
rattlensnake master, 78
recipes, 134, 135, *135*
recycling, 18, 180
red autumn fern, 30, *31*
red-leafed bromeliad, 22, *23*
redwood, *62, 63, 63*
relaxation, *146,* 147
retreat, *155,* 155–57, *156, 157*
Riccio, Peggy, 130
Rice, Valerie, *133,* 133–34
rocks, *47, 48–49,* 100, *100–101*
Roger's Gardens, 192–93, *193*
root system, 75, 134

rosette, *44, 46, 47*
row covers, 123, *124–25*
Royal Horticultural Society, 50

S
saguaro, *16,* 22
Salman, David, 47, *48–49*
Sargaent, Juliet, 19, 20
Schmidt, J. Frank, 55, 56
Scott, Lori, 40, *40*
seaside daises, *70–71,* 71
seasonality, 45, 50
Seattle, 80, *81,* 102, 106, *106–7,*
    182
sedge meadow, 14, *14*
seeds, *116,* 117, 118–20, *119–20*
    saving, 140
    varieties of, 122–23
shelter, 154, *154*
Shepherd, Rene, 104
short-season, 122–23
showy milkweed, *39*
shrubs, 14, *14, 52,* 53, *57–59, 58,*
    167
    native, 173, *173*
side yard, 30, *31,* 32
silver echeveria succulent, 93, *93*
silver leaf mullein, 32
Sluis, Janet, *68,* 68–69
soil, 92–93, 109, 134
    micro-organisms in, 85–89, *87*
soil biota, 86–89, *87*
Sonoran Desert, *16,* 22
space, 32, 150, *151,* 154, *154,* 162
Speight, Teresa, *178,* 178–80, *179*
sporotrichosis, 147
springtime, 153, *153*
stewardship organizations, 193
St. John, BJ, 111, *111*
St. John, Primus, 111, *111*
stones, *52,* 53, 100, *100–101*
strawberry tree, *39*
stumpery, 103, *103*
succulents, 42–44, *43,* 166, *166*
Sunset Western Garden Book
    (Vanderhoff), 193

Sussex, England, 20, *20*
sustainability, 10, 18, 86, 87, *87,*
    106, *106–7*
sword ferns, 103, *103*

T
Tallamy, Doug, 177
A Tapestry Garden (O'Byrne, E.,
    and O'Byrne, M.), 61
Tasmanian tree fern, 103, *103*
therapy, horticulture, *160,* 160–62,
    161
Thomas Jefferson University,
    182
tomatoes, 130, 131, *131,* 134, *138,*
    139
Tongue of Fire (Borlotto Lingua di
    Fuoco), 94
tools, 90, *90,* 144, 150
tranquility, 151–53, *152, 153*
transition zone, 28, *28*
trees, 54, 55, 56, 195. *See specific*
    trees
    allergies from, 190
    deciduous, 167
    fruit-bearing, 172
    shade for, 57, 188–89, *189*
    street, 187, *190,* 190–91, *191*
trellises, *116,* 117, 118, *119*
Tucson, Arizona, *16,* 22
Turkish stonecress, 47, *48–49*
Turner, Mark, 130

U
UK Centre for Ecology and
    Hydrology, 50
Urban and Community Forestry
    Department, 55
urban farm, 106, *106–7*
urban garden, 181
urn fountain, 100, *100–101*
USDA Forest Service, 55

V
Vancouver, 164, *164*
Vanderhoff, Ron, 192–93, *193*

Vashon Island, 165, *165*
vegetable
    foodscaping and, 112–14, *113*
    garden, 104–11, *105*, *106–7*, *108*,
        110, *111*
Verbena bonariensis, 175, *175*
Vermont, 195
veronica, *70–71*, 71

W
warm-season, 118, 121
Washington, DC, 130

Washington State, *52*, 53, *110*, 111,
    122–23, 165, *165*
waste, 89, 91
water, 41, 109
    within garden, *163*, 163–65, *164*,
        165
    garden on, 92–95, *93*, *94*, *95*
weeds, 72, 85
white gaura, 95, *95*
Wiggins, Charlotte Ekker, 102
wildflower, *26–27*, 27
Wildflowers of Orange County

(Vanderhoff), 193
wild lands, 194, *194*, 195
wildlife, 17, 73, *73*, 175, *175*
wild quinine, 77, *77*
wild rye, 173, *173*
wildscaping, 173, *173*
window box, 80, *80*
Windy City Harvest, 183, *183*
Wisconsin, 76, *76–78*, *77*
work, 148–50, *149*
wormwood, 47, *48–49*
Wynton, Marina, *82–83*, 83

Photograph Credits

Brie Arthur: p. 113; **Caitlin Atkinson:** p. 68; **Rob Cardillo:** pp. 1, 23, 35–36, 152–153, 156, 175, 197; **Jennifer Cheung:** pp. 146, 168, 169; **Chicago Botanic Garden:** pp. 160, 161; **Robin Cushman:** pp. 73, 87, 90; **Darcy Daniels:** p. 31; **Tom Grey:** p. 11; **Caroline Greyshock:** pp. 13, 163; **Sara Hall:** p. 193; **Saxon Holt:** pp. 3–4, 5–6, 14–15, 26–27, 28–29, 39, 43, 46, 47–48, 70–71, 97, 99, 100–101, 105, 106–107, 108, 126, 127, 128, 131, 132, 141, 151, 173, 176, 194, 189; **Heidi Hornberger:** p. 149; **the Ingalls:** pp. 133, 135, 136–137; **Bonnie Kittle:** p. 202; **Jeff Lafrenz:** p. 54; **Jason Lewis/Los Angeles Standard Newspaper:** p. 189; **Mary-Kate Mackey:** pp. 63, 179; **Melinda Myers:** p. 77; **Matt Thomas:** p. 20; **Mark Turner:** pp. 124–125, 183, 185, 190, 191; **Rachel Weill:** pp. 2, 8, 84, 115, 116, 119, 120, 138-139, 142, 145, 170, 171; **Doreen Wynja:** pp. 16, 25, 34, 40, 52, 55, 58, 60, 62, 65, 66, 80, 81, 82-83, 92, 93, 94, 95, 103, 110, 111, 154, 155, 157, 158, 164, 165, 166, 181

Editor: Shawna Mullen
Designer: Darilyn Lowe Carnes
Managing Editor: Lisa Silverman
Production Manager: Rachael Marks

Library of Congress Control Number: 2021932533

ISBN: 978-1-4197-5461-6
eISBN: 978-1-64700-287-9

Abrams books are available at special discounts when
purchased in quantity for premiums and promotions as well
as fundraising or educational use. Special editions can
also be created to specification. For details, contact
specialsales@abramsbooks.com or the address below.

Abrams® is a registered trademark of Harry N. Abrams, Inc.

ABRAMS The Art of Books
195 Broadway, New York, NY 10007
abramsbooks.com

Page 1: **A blooming bee magnet.** A honeybee draws nectar from a sea holly (*Eryngium amethystinum*). This perennial's showy bracts grow in clusters atop stems 2½ feet tall.

Page 2: **Entice butterflies.** Choose plants whose blooms produce nectar and you might even encourage them to stay awhile—especially if your garden provides what they need (shelter, water, and nectar). Pictured is a rare pipevine swallowtail butterfly, pausing on a flowering currant (*Ribes sanguineum*).

Pages 4–5: **Connect with nature.** If possible, locate your gathering place where you will feel like you're on vacation. This patio, off the side of a house near Charlottesville, Virginia, blends nicely into the forest beyond. Rustic bentwood chairs complement the gardens' woodland feel; snow azaleas bloom behind.

Pages 6–7: **Raised beds to the rescue.** Raised beds occupy much of a front yard. Because they sit atop the ground, they provide ideal conditions for growing food, including perfect drainage. Choose cedar or redwood to build them; it lasts longer than other woods. For easy harvest, make your beds about eight to ten feet long, and four feet wide. Foil gophers by underlining the beds with hardware cloth.